Clare Tipler

NEITHER GHOST NOR MACHINE

NEITHER GHOST NOR MACHINE

THE EMERGENCE AND NATURE OF SELVES

JEREMY SHERMAN

Columbia University Press
New York

Columbia University Press
Publishers Since 1893
New York Chichester, West Sussex
cup.columbia.edu

ISBN 978-0-231-17332-2 (cloth)
ISBN 978-0-231-17333-9 (paper)
ISBN 978-0-231-54599-0 (ebook)

The Library of Congress has cataloged this record under
LCCN: 2017028428.

Columbia University Press books are printed on permanent
and durable acid-free paper.

Printed in the United States of America

FOR TERRENCE W. DEACON

CONTENTS

FOREWORD

TERRENCE DEACON

When longtime colleague Jeremy Sherman first approached me with the idea of producing a brief and simplified account of the theory presented in my six-hundred-plus-page book *Incomplete Nature* I was quite skeptical. More than one academic reviewer has failed to grasp its central theme, and many have fallen prey to the expedient of assimilating it to currently popular paradigms that it instead critiques. Readers have often commented that the density and diversity of subject areas the theory covers make the main ideas difficult to assimilate. Yet others have worked to create road maps into the material to help readers see the core paradigm-challenging claims being offered, but with only partial success. Even I have found it impossible to compose the "elevator speech" that summarizes the main ideas.

So I have assumed that an effort to present its most important ideas to an audience with no particular scientific or philosophical preparation is unlikely to succeed. Following Einstein's rule of presenting an idea as simply as possible but not too simply, wouldn't such a distillation be too simple?

Neither Ghost nor Machine has begun to change my mind about this.

I have known and worked with the author for nearly two decades discussing these ideas. We have coauthored a few short papers presenting some of them and his input has played an important role in fine-tuning many of the ideas developed in *Incomplete Nature*. While I have been producing books and papers for an academic audience, he has been a prolific blogger, producing a widely read blog for *Psychology Today*. So, if anyone could communicate these ideas to a lay audience, he should be able to.

Over the course of more than a year, we have had dozens of conversations about ways to approach this material. During the process of these discussions, one of the main aims was to determine how to present the theory with the fewest technical terms and minimal scientific details, while still communicating the core ideas. Chief among these aims was finding common terms that convey the central concepts most accurately, but in a way that makes them seem familiar while at the same time highlighting critical unquestioned assumptions about their meanings. In these deliberations, it became clear that the core concepts can be exemplified by the commonsense notions of *selves* and *aims*.

Though everyone is familiar with these concepts and uses them daily without a second thought to explain what goes on, they are not so innocuous when they are imported into the natural sciences. Indeed, they are all but forbidden because of the ways they often serve only to masquerade as explanations. But selves and their aims aren't illusions. Human and nonhuman aims have radically altered the surface of the planet. It is, therefore, bordering on the absurd that our current "theories of everything," which purport to provide the most fundamental explanations for all that exists, should simply fail to include an explanation for the very properties that theorizing itself depends on.

For some reason, we just don't seem to have well understood commonsense scientific concepts to handle these sorts of phenomena. This helps to explain why two of the most enigmatic scientific mysteries of our age are the nature of a conscious self and the origin of life. I have argued that it is not the technical complexity as much as the counterintuitive nature of these phenomena that is the problem. So, the task of producing something like a "beginner's guide" to solving these mysteries, without delving into scientific detail or introducing esoteric new terms and concepts, is a daunting one. Can this be done simply but not too simply? Can the analogies used to provide insight into the essential principles avoid misinformation and yet convey essential insights that have so far evaded our best science? Can the use of commonsense language convey the essence of concepts that are in many respects quite alien to common sense?

As I have witnessed the gestation of this book through dozens of rewrites and edits, I have been impressed by the care taken to find just the right terms and examples. The choice of *self* and *aims* to convey the core ideas in

Incomplete Nature instead of the neologisms *autogenesis* and *teleodynamics* is a great example of such a choice. But so is the distinction made between *regularization* and *self-regeneration*, as they are used to capture the difference between, for example, whirlpools and organisms, respectively. Even where the same molecular thought experiment described in *Incomplete Nature*—an autogen—is used to exemplify the transition from inanimate to animate (functionless to functional) systems, the account makes use of metaphors and analogies that make it seem familiar and imaginable. The result is a solid first step toward making the unfamiliar familiar and the esoteric relevant.

But this book is more than merely a simplified précis of *Incomplete Nature*. The larger context of my research, which motivated me to explore these issues in the first place, includes decades of neuroscience research and an interest in the evolutionary process that produced such distinctive human capacities as language and symbolic reasoning. *Neither Ghost nor Machine* makes connection with some of this work. In particular, it shows how this account of selves and aims can help to explain the surprising role played by relaxation of selection in the evolution of biological complexity and higher-order cooperative behaviors.

Finally, as this foreword might demonstrate, my style of writing is perhaps too steeped in the academic tradition to be easily assimilated by the general reader. And besides, only the most dedicated reader can slog through the over six hundred pages of *Incomplete Nature* required to get the full story. In contrast, *Neither Ghost nor Machine* is a brief read, written in an accessible, conversational tone that won't require periodic rereading of contorted sentences to get their gist or running to the dictionary or the glossary to interpret unfamiliar terms.

Whether as a stepping-stone to reading more technical books and papers on these topics, a means to sweep away a few of the cobwebs of intellectual complacency, or an exercise in thinking a few radical thoughts on topics encompassing all of what matters to us most, this book will be sure to reward the curious and open mind.

NEITHER GHOST NOR MACHINE

I

OVERVIEW

1

THE MYSTERY OF PURPOSE

WHAT ARE WE?

Every generation marvels at what prior generations didn't know, the mysteries they hadn't yet solved, perhaps hadn't even noticed. We might, therefore, wonder what future generations will look back at as our biggest blind spot. What central scientific mystery haven't we solved yet and perhaps haven't even noticed?

This book is about a likely candidate, the mystery of purpose:

What is purpose and how does it emerge from purposeless phenomena?

It also proposes a heretofore-unexplored natural science solution to the mystery.

While the term *purpose* often refers to deliberate, conscious, intended, or declared goal setting, here it will include all traits and behaviors that are functional, valuable, significant, or useful, including all biologically adaptive traits. To expand beyond the narrower implications of the term *purpose*, I will mostly refer to purposes as aims. Thus, the mystery of purpose becomes how aims emerge from aimlessness.

Purposefulness or aiming is unimaginable without reference to agents, organisms, individuals, or beings. For example, when biologists say that adaptations are functional or serve purposes, they can't help but identify the organisms whose purposes are served or the aims that those organisms express.

Here I will refer to beings, organisms, individuals, or agents as *selves*. Selves have purposes or aims. Thus the mystery of purpose becomes how selves and aims emerge from self-less, aimless physics and chemistry.

Physics and chemistry are cause-and-effect phenomena. In contrast, selves and aims refer to means-to-ends behavior. Means-to-ends behavior depends on but is not reducible to cause-and-effect phenomena. Despite great effort over centuries to explain away means-to-ends behavior as nothing but cause and effect phenomena, here I'll argue that it can't be done, and that there is no alternative to explaining a kind of phase transition whereby means-to-ends behavior emerges from cause-and-effect events—not just that it must have or that it could have, but how it really does here or anywhere in the universe where means-to-ends behavior exists.

We do not yet have a solution to the mystery of purpose, but we may not have to wait generations for one. This book presents an unprecedented path to the solution, one that doesn't explain selves and aims as the product of phantom ghostlike forces, or explain them away as nothing but cause-and-effect mechanisms. Selves with aims are neither ghosts nor machines, and yet they are strictly natural; their special properties have a perfectly feasible explanation within the known laws of classical physics and chemistry.

FOUR QUESTIONS

The mystery of purpose is one huge mystery in four questions:

The Nature of Selves: What distinguishes selves from nonselves? What, for example, distinguishes you from lifeless chemistry or a computer?

The Origin of Selves: How could selves emerge in a universe that apparently didn't contain them at its beginnings?

The Nature of Aims: What distinguishes self-directed, means-to-ends behavior from cause-and-effect phenomena? For example, what makes you strive to survive and thrive, pursuing what matters to you? What does all functional, adaptive, useful, good, helpful, beneficial, purposeful, intentional behavior have in common and what makes such behavior distinct from things that just happen unaimed, unintended, and without purpose in lifeless physics and chemistry?

The Origin of Aims: How could aims emerge in an otherwise aimless universe? In other words, how does mattering emerge from matter?

These four questions fold into one big mystery for two reasons.

1. Selves and aims are inextricably linked: Only selves aim. To be a self is to aim; to aim is to work toward some potential ends, and the most fundamental such state is to remain a self. Nonselves have no aims of their own. We selves build machines to serve our aims, not their aims. Nothing matters except to selves, given their aims.
2. Origins and natures are inextricably linked: We can't tell how something emerges if we don't know what emerges, and we can't know exactly what has emerged without knowing how it emerged. For example, we know that we are selves, but without knowing how selves emerged, we can't really know what it means to be a self. Without a solution to the mystery of purpose, we don't yet know what we are.

DISMISSING THE MYSTERY

Like many big mysteries unsolved by past generations, the mystery of purpose is one that many people hardly consider worth solving. When asked about it, scientists tend to respond as though it's irrelevant to their interests. They will tend to shift attention toward other questions that to them are more familiar and worthy of their attention.

The mystery of purpose has ancient roots. In the divorce settlement three hundred years ago, when natural philosophy separated from philosophy, taking the name natural science, philosophy got to keep the mystery of purpose. Natural scientists expected that science would get along just fine without having a solution to it.

In her deep intellectual history of debate about living agency, historian Jessica Riskin offers a glimpse of how science is fairing without a solution to the mystery of purpose. She describes a biologist friend, who notes that

> It is absolutely against the rules in her field to attribute agency to a natural entity such as, say, a cell or a molecule, but she also agreed that biologists

> do it constantly, just as a manner of speaking: they speak and write as if natural entities expressed all sorts of purposes and intentions, but they don't mean it literally. "Sure, we do it all the time, when we're teaching, in lectures, even in published articles. But it's just a sort of placeholder for things we don't know yet. The more we get to know, the less the phenomena will seem purposeful.[1]

The taboos this biologist describes are ambiguous, expressing unresolved ambivalence endemic in the life sciences, leading to all sorts of equivocation. According to Riskin's biologist friend:

> Certain verbs are worse than others: those that seem "anthropomorphizing," such as "want," are only permissible in casual settings. Biologists can say, and allow their doctoral students to say, that "cells want to move toward the wound" in conversation but never in print. In contrast, other active verbs do not seem anthropomorphizing. . . . Proteins "control" chemical reactions; muscle cells "harvest" energy; genes "dictate" the production of enzymes.[2]

The "central problem of biology" as Nobel Prize–winning French biochemist Jacques Monod put it, is "how could purposeful systems have emerged from a universe with no purpose?"[3] One nonanswer is to simply declare that it emerged, without saying how. Reviewing scientific responses, biochemist Addy Pross said, "the minimal attention that has been directed toward this 'central problem' suggests that the scientific community considers the problem solved (or uninteresting) and has accepted the 'emergent property' explanation."[4]

Thus, one can sidestep by begging the question "How did purposeful systems emerge?" "By emergence!" One can also sidestep it by defending equivocation. To give but one example of many, consider physicist Sean Carroll's argument from his best seller, *The Big Picture*: "Those swirls in the cream mixing into the coffee? That's us. Ephemeral patterns of complexity, riding a wave of increasing entropy from simple beginnings to a simple end. We should enjoy the ride."[5]

This might make us wonder whether swirls of cream enjoy the ride. The question doesn't occur to Carroll, who defends equivocation by promoting an approach he calls poetic naturalism as the argument that "there is only one, unified, physical world, but many useful ways of talking about it, each of which captures an element of reality. Poetic naturalism is at least consistent with its own standards: it tries to provide the most useful way of talking about the world we have."[6]

To Carroll, "The appearance of something like 'purpose' simply comes down to the question 'Is "purpose" a useful concept when developing an effective theory of this part of reality in this particular domain of applicability?'"[7] And elsewhere, "The idea that something wants something else is a way of talking that is potentially useful in the right circumstances—a simple idea that summarizes a good amount of complex behavior in a convenient way."[8]

To paraphrase, according to poetic naturalism, we can say that selves want if we, as selves, want to. Of course this is true of everyday explanation, but scientists need to hold a tighter standard. It's not enough to explain purposeful consequences by means of other purposeful consequences—we want to assume wants. We have to address Monod's central question, explaining how purposeful consequences emerge from nonpurposeful consequences, or in Monod's words, how purposeful systems have emerged from a universe with no purpose.

NOT JUST HUMAN SELVES

The first selves to emerge from aimless physics were not humans. Our species appeared on the earth only moments ago in the history of life. So just how far down the food chain should we expect to find selves? Realistically, we need to go all the way down to the simplest known organisms and indeed further still, to the first possible organism here or anywhere else in the universe in order to explain the emergence of selves and aims.

People tend to think of selves and aims as distinctly psychological phenomena—selves as self-awareness; aims as conscious intentions or stated goals and purposes. The mystery of purpose runs much deeper than that.

To corner ourselves with the mystery, here I'll take the radically inclusive step of regarding every known and unknown organism as a self. By my definition, even trees, the simplest microbes, and as-yet undiscovered extraterrestrial life forms are selves. Every living being that ever existed or ever will exist here or elsewhere, from its conception to its death, is a self. To solve the mystery of purpose requires explaining how selves and aims could emerge anew anywhere in the universe.

Offsetting my inclusive definition of selves, I'll be excluding some selves commonly assumed to solve the mystery. First, I'll exclude any supernatural selves. This includes gods, higher powers, souls, spirits, and any nonphysical élan vital or universal life force. Scientists seek explanations within the natural, not the supernatural, realm, which by definition is beyond nature and therefore beyond yielding any empirical evidence whatsoever. Besides, animating inanimate things is altogether too easy. Any of us can imagine Gods as selves with aims as easily as we can imagine that rocks aim to fall, rivers strive for the sea, or the whole universe wants us all to be here. This makes for beautiful, evocative poetry, mythology, spirituality, religion, and fable, but it won't work for science.

I'll also exclude natural selection, DNA, and RNA—the chemicals of life. These are not selves and they have no aims.

There's a tendency, even among some biologists, to talk about natural selection as a purposeful self, aiming to design and improve organisms. The term *natural selection* encourages this false impression—Mother Nature selecting organisms that satisfy her aims.

Current popularizations of Darwinian theory also encourage the false impression that natural selection has aims. Darwin argued that evolution occurs through the interplay of heritability, variation, and selection. Many popularizers of evolutionary theory have oversimplified heritability and variation down to purposeless replication, the imperfect copying of inanimate molecules. If that were all there was to heritability and variation, then life's aims could be explained by natural selection aiming to design organisms or program DNA or by inanimate replicating chemicals aiming to self-replicate.

However, there is a difference between chemical replication and what selves do. Replication is the mere proliferation of inanimate molecules. Most chemical reactions yield a proliferation of molecular products. But chemical reactions are ephemeral due to the second law of thermodynamics—the universal tendency for organization to become disorganized, generated concentrations to degenerate, and energy to dissipate. Chemical reactions peter out.

Purpose is not found in chemical reactions, not even in prebiotic proliferation of DNA or RNA, the molecules that now play crucial roles in all known forms of life.

Though molecular replication eventually peters out, selves have persisted for billions of years. Selves are *self-regenerative* in two senses: they maintain their own existence, and they produce new selves. They somehow have a capacity to outpace the second law, regenerating themselves faster than they would otherwise degenerate. Selves don't just peter out. If they did, lineages of selves would not have lasted for billions of years.

While we can imagine the differential replication of molecules as a little like evolution, biological evolution is different, and must have begun with the emergence of a capacity for *self-regeneration*, not merely chemical replication. To solve the mystery of purpose will require explaining how self-regeneration emerges from chemistry.

While natural selection hones populations of selves and aims, it doesn't create them. To claim that natural selection explains purpose is like claiming that erosion explains mountains. Erosion, also a second-law process of degeneration, explains how mountains are passively sculpted, but not what's sculpted. Likewise, natural selection explains how populations of selves are passively sculpted through differential reproductive success, some lineages producing more offspring than others, but not how selves arise in the first place.

People concede that natural selection doesn't solve the mystery of purpose whenever acknowledging that, although we have a mature science of evolution, the origin of life remains a mystery. Yet we often seem to forget this, treating evolutionary theory as though it were a complete explanation for life.

So by the definition of selves used here, protozoa are selves but natural selection and replicating chemicals are not and DNA and RNA are molecules, not selves. They have no aims, selfish or otherwise.

If we are to solve the mystery of how selves and aims emerged, we must explain selves prior to evolution, evolvable selves that didn't themselves evolve, selves that came into existence by accident from mere physics and chemistry.

WHEN PURPOSES EMERGE

Selves are self-directed with all that the term implies: Selves demonstrate autonomous initiative or agency. To some extent, they have the ability to organize themselves and are under their own control.

The term *self-directed* also suggests *serving oneself*, doing work directed toward self-benefit. In contrast to molecules, and morality aside, selves are *selfish*, meaning somewhat self-serving, in the same way that *bluish* means somewhat blue. Selves aim, organize, or focus work for their own benefit, doing self-regenerative work *by themselves for themselves*.

Physical work is common in the inanimate world, while self-directed work is rare. Water running downhill does physical work when it carves canyons. Wind does physical work when it blows leaves off tree branches. But blowing wind and running water don't do this work for their, or anything else's, benefit. This work just happens.

Machines perform focused or aimed work, but not for their own benefit. They serve the aims of the selves who design and use them. Only selves focus or aim work for their own benefit. Selves work to serve themselves because they are in constant flux and their fragile organization would otherwise naturally degenerate. Realistically, then, a viable solution to the mystery of purpose would explain the emergence of self-directed, self-regenerative work.

As explored later, the second law of thermodynamics explains why chemical interaction peters out, and why perpetual motion machines are impossible. Of course, selves aren't perpetual motion machines, but they are self-perpetuating, lineages of selves sustained over billions of years because they channel energy into self-regenerative work, which includes, but is not limited to, *self-reproduction*, the generation of offspring that inherit the self's capacity to do self-regenerative work.

Wherever and whenever selves and aims emerge anywhere in the aimless universe, they must channel work into self-regeneration to outpace the second law. Like the Red Queen in *Alice Through the Looking Glass*, selves have to "run" to stay in place. Their focused work is aimed at not falling apart, not dying, not succumbing to the second law. This is the self's defining purpose.

When selves first appear anywhere in the otherwise aimless universe, self-regeneration emerges also. Since chemistry has no aims, purposes, or intentions, the first purpose or aim—self-regeneration—must emerge by accident, not "on purpose."

Creation myths about the origin of purpose sidestep physical and chemical challenges that can't be sidestepped in a scientific solution to the mystery of purpose. Abandoning supernatural explanation, scientists must seek the origin of purpose in the only selves for which there is scientific evidence: locally emergent, self-serving selves that are fragile, fallible, at risk, their self-directed work hit or miss.

Considering more closely the meanings of the three words in the term *self-directed work*, we can define what a solution to the mystery of purpose must explain:

Work: Spontaneous, inorganic, physical work is neither directed nor self-directed. For example, the sun's radiation striking the Earth does work when it heats the ground, but the work is not directed to have this effect, nor does this work serve the sun's purposes or aims. The sun doesn't regenerate itself through its work, but instead, like all stars, is in the process of burning up its nuclear "fuel" as fast as possible.

Directed: Many processes do work that is directed, but not *self*-directed. For example, a machine such as a computer performs highly focused or directed work, but not to serve the computer's purposes. Like the work performed by an automobile engine or clock, the work of computation is directed to produce a specifically focused result determined by a user or designer of that mechanism—a self.

Self: Selves and only selves do self-directed work.

STUBBORNLY MYSTERIOUS

In the last few centuries, scientists have solved a great many of the mysteries relevant to the temporal and spatial scales at which selves and aims exist. With regard to biochemistry, for example, we have identified the chemicals that bodies are made of, and we understand an enormous amount about the cause-and-effect interactions between those chemicals.

The life and social sciences have likewise flourished. From biology to sociology, we now have a substantial and growing capacity to describe the functional means-to-ends behavior of the simplest organisms and the largest human societies.

You would think, therefore, that with all of this light shed on living behavior from opposite sides—the cause-and-effect realm of the physical sciences and the means-to-ends realm of the life and behavioral sciences—the seam where these two realms connect would now be illuminated.

It isn't. We still can't say how mattering emerges from matter, means-to-ends from cause-and-effect, selves from chemistry, purposes from purposelessness, functional behavior from functionless events or aims from aimlessness.

Here we face a mystery at the very heart of life, one that is likely to involve only our most reliable and thoroughly understood physics and chemistry, without recourse to the big bang, quantum mechanics, black holes, or dark matter. Yet, we have no solution to the mystery of purpose.

About nothing so central to our lives do we remain so deeply in the dark.

TELEOLOGY

The mystery of purpose is not new. It has taken many forms over the millennia of human inquiry, though rarely framed as bluntly as I mean to pose it here. In theology, the mystery is addressed through questions about the nature of spirit, soul, or God, who is taken to be a primordial self with all-encompassing aims.

In philosophy, the mystery is approached from many angles, including attempts to explain the nature of mind, knowledge, love, will, morality, and values, all qualities we associate with selves, though especially human ones. Indeed, philosophers have tended strongly to assume that these properties start with humans and rarely wonder about their emergence in the universe.

In academic discussion, the mystery of purpose is explored as *teleology* (*telos* is Greek for purpose). Teleology is most strongly associated with theology, where a selflike God's purposes and aims are the focus. As we'll see later, during the Enlightenment, teleology became something of a philosophical backwater as scientists turned their attention toward cause-and-effect explanations for everything.

Scientists have expressed ambivalence about addressing the mystery of purpose, sometimes treating it as outside the scientific purview, sometimes treating it as already or soon to be solved or dissolved by scientific discovery. The mystery surfaces as scientific whenever a new artifact suggests a solution by analogy, such as the self as clockwork or the mind as computer—or whenever a newly discovered force or phenomenon suggests a solution, again by analogy, from Mesmer's animal magnetism to Freud's libidinal pressures to, today, "quantum consciousness." As we'll see, these analogies don't solve the mystery of purpose.

SOLVABLE

This book presents a new solution to the mystery developed by University of California, Berkeley, scientist Terrence Deacon with the support of others, including myself. Deacon is a Harvard-trained neuroscientist and biological anthropologist who has done important work on the evolution of the human brain and language before tackling the mystery.[9]

Rather than starting with human selves and their aims or even the simplest known organisms, Deacon decided that he had to explain how the very first selves could emerge anywhere in the universe; in other words, he sought the missing link from aimless phenomena to selves and aims. He has developed an empirically testable proof of concept demonstrating that it is possible within basic physical laws for true selves and aims to emerge.

His approach may seem like origin-of-life research, but it's not. Origin-of-life research tends to sidestep the transition from cause-and-effect events to means-to-ends behavior. For example, the assumption behind much origin-of-life research today is that once RNA molecules started replicating aimlessly, they could be honed by natural selection, which is also aimless.

Within most origin-of-life research, life is a transition within cause-and-effect events, not a transition from cause and effect to selves engaged in their own means-and-ends behavior. As such, the mystery of purpose remains unsolved, even unaddressed.

In contrast, Deacon has focused specifically on the transition from cause and effect to means and ends, an exclusively natural transition from non-selves to selves, aimlessness to aims, purposelessness to purposes. To do so, he has assumed an absolutely sterile environment, a universe with no selves or aims—no will, no natural selection, not even some scale-tipping natural tendency for selves and aims to emerge. In this sterile context, he has found a way that natural cause-and-effect chemical processes could fall into the self-directed, means-to-ends work evident in real selves with real aims.

Entering into this research, he assumed somewhat paradoxically that the process by which selves and aims emerged must have been simple. With nothing and no one to evolve or engineer the first selves, and life playing out at the scale of classical physics and basic chemistry, it can't have been that complicated.

PROCESSES OF ELIMINATION

Deacon guessed that the solution has eluded us for so long because we have been making tacit assumptions that get in our way. These assumptions are not consistent with current science but have been carried over into it as vestiges of past approaches or human intuition. The most fundamental of these assumptions is that all changes are produced *positively* by the addition of things or forces, for example, that there must be something added to matter to cause it to come alive.

In recent centuries, scientists have come to recognize that changes can result *negatively*, through *processes of elimination* not of things, but of possible dynamic paths. *Dynamics* are the interactions throughout large populations of things. For example, consider water molecules flowing in various currents. The currents can get in one another's way, creating impasses that reduce the likelihood of water molecules moving down some paths compared to others. In other words, *constraints* can emerge through dynamic interaction.

As I'll show later, major insights, including those that spawned self-organization, evolutionary, and information theories, were born of the recognition that emergent constraints can make some possible paths less likely, thereby making other possible paths more likely and resulting in fundamental changes in character.

We often say of synergy that the "whole is greater than the sum of its parts." Deacon turns this on its head with the declaration that "the whole is less than the sum of its parts." This will take some unpacking, but one implication is that life is not something added to physics and chemistry, but rather a reduction in physical possibilities that emerges through dynamic interaction.

Take the molecules that compose your body and consider the vast number of ways they could interact. Now think of how few of those ways are possible within the living self you are. There's what's possible in physics and chemistry, and there's what's possible in selves, and the possibilities within selves are less, not more. A dead body's materials can be in vastly more arrangements than a living body's materials.

With selves, nothing is added, nor is a greater quantity magically produced through synergistic combination. Rather, when things interact dynamically, possible paths of interaction are subtracted or eliminated through a process somewhat like gridlock—paths blocking paths. According to Deacon, selves and aims are the result of emergent dynamic constraints: not something added, but possible paths subtracted. Self-regeneration, the self's first purpose or aim, is made possible by emergent constraint, a reduction in the likelihood of paths that are not conducive to self-regeneration.

CONSTRAINED SELF-REGENERATION

Self-regeneration is a circular, looped, or iterative capacity unique to selves. Being alive, selves work to stay alive, doing work that we are able to do because we are alive. Our means and ends are circular. We engage in means-to-ends behavior most fundamentally toward the end of maintaining our self-regenerative means.

By this account, selves and aims originate as one and the same. The self is the aim to self-regenerate; the aim to self-regenerate is the difference between a living self and a dead body.

Self-regeneration describes the added capacity found in selves, but not how it's achieved. Indeed, the focus on self-regeneration as the self's added capability distracts from how the capacity is added. It's added through subtraction, a negative process of elimination that reduces the likelihood of self-degeneration.

To solve the mystery, Deacon needed to identify the processes of elimination that would keep self-regeneration happening. It would do so by limiting the tendency for self-regenerative dynamics to fly out of control, ending the circular self-regeneration of self-regeneration necessary for life.

To explain the emergence of purpose—of selves and aims—we, therefore, must explain how, through interactions, molecules could ever fall into a kind of circular trap, a constraint on the tendency to degenerate that regenerates that selfsame constraint.

I have studied and collaborated with Deacon for twenty years. Here I present my interpretation of his theory as simply as I can to provide a feel for his project and its implications as I see them. In places, I have extrapolated beyond his published research and added ideas of my own. Deacon has reviewed and critiqued this book in detail. Still, I take full responsibility for any misinterpretations of his theory.

I'll provide a preview of his solution, but before I do, let's bring the mystery into a little more focus because it's an oddly elusive one, right in front of our noses but rarely on the tip of our tongues.

2

THE BIGGEST MYSTERY WE EVER IGNORE

CLOSE, BUT NOT PRESSING

Few of us give much attention to how selves emerge and, with them, functions, purposes, values, meanings, intentions, and significance—in a word, aims.

Though the mystery of purpose is not on the tip of our tongues, it lurks behind the everyday questions that are. We spend our lives achieving aims but also, in moments of uncertainty, wondering what the right aims are.

Day to day, we don't typically seek the meaning of life. We mostly seek ways to make our personal lives meaningful. It's humankind's search for a little meaning, just enough to stay motivated. Our practical quest for appropriate personal aims—right friends, partner, work, beliefs, and the like—rarely corners us with the bigger question: What are aims anyway?

Likewise, our selfhood is a lifelong preoccupation. We attend to our self-image, status, appearance, value, standing, opinions, and deeds. Caring as much as we do about ourselves, most of us have a very hard time imagining our selfhood vanishing. Little seems more solid to us than the selves we are, or more worth protecting and maintaining. Still, this preoccupation with our individual selfhood does not carry us all the way to the question of what a self actually is.

Way off on the horizon of our personal experience we glimpse the mystery of purpose, big and blurry, not pressing, not a mystery we need to solve in order to get on with our lives. We might think it's unsolvable, or maybe solved already, at least abstractly, with talk of souls, energy, creators, higher powers, evolution, or chemical mechanisms.

We might sense that these are incomplete or inadequate answers. We might favor one and scorn the others as laughably inadequate, but in doing so, we don't usually bother to detail how our favored solutions actually work. What created the creator? How is vital energy or a higher power different from electricity? What is a soul made of? How does evolution give rise to life?

Day to day, it's not urgent that we solve the mystery. Unsolved for millennia, it taunts us elusively, and we taunt back, ridiculing "the meaning of life" as a navel-gazing question best ignored if we are to be productive selves achieving our most pressing aims.

BLURRING THE DIFFERENCE BETWEEN CAUSE AND EFFECT AND MEANS AND ENDS

We easily distinguish selves from aimless matter—a live chicken versus a fried one, a seed versus a grain of sand, a tree versus a stalagmite, a working brain versus a computer. But what accounts for this difference?

In everyday life, we assume that some things are inanimate objects and other things are animate selves, even though we can't quite identify the critical difference. We say magnets pull, but people want; the wind blows a door open without any intention of doing so, but the dog paws the door open because it aims to get in.

We also manage by means of loose metaphors that enable us to blur the distinction between causes and means, effects and ends. We animate cause and effect; we deanimate means and ends. Animating, we say that the wind craves entry but the door is stubborn, or that the malfunctioning computer aims to thwart our effort to achieve our aims. Deanimating, we talk about ideas pushing and pulling others to our way of thinking or being attracted to people because of good chemistry.

For all our progress in the physical and life sciences, we don't yet have a way to bridge from one to the other; we have no way to integrate while still distinguishing the realms of purposeless and purposeful phenomena.

The physical sciences can explain what happens in their realms exclusively in terms of cause and effect, but the life sciences can't account for behavior as cause-and-effect phenomena. Organisms aim. Organs function. Brains interpret. Even if just by unspoken implication, the life sciences cannot explain what happens in their realms without means-to-ends concepts.

If a physics professor told you the moon pulls the tides in order to achieve some aim, you would raise a skeptical eyebrow. But you're perfectly comfortable with a biologist's claim that the simplest organism engages in functional, aimed behavior, or with a social scientist's claim that purposes determine many human actions.

Running down the halls of academic and scientific institutions there's an invisible line we all know not to cross, the line between what can and can't, must and mustn't, be explained in terms of selves and their aims. We honor the line. We teach our children about it. They might initially mistake moons as wanting, but they soon outgrow their trespasses.

Despite our use of animating and deanimating figures of speech, the line is, for the most part, well drawn, except in our poetic and fanciful abstractions. We respect the line; we just haven't yet explained it.

SCIENCE BY EQUIVOCATION

We tolerate blurring by animation and deanimation in everyday life, but not only there. Blurring is tolerated to some extent in the sciences in an approach I'll call *equivocation*, using a term in two senses without attending to important differences of meaning.

Natural selection is a good example of this blurring. The term can be employed to refer to the physical effects of the inanimate environment on organism reproduction, or figuratively as an animate self, natural selection aiming to *design* traits.

Likewise, all means-to-ends motives can be treated equivocally as inanimate causes of effects or as animate selves. For example, an appetite or drive might in some context be counted as the secretion of hormones or neurochemicals and in another context as a little person in our heads saying, "Go for it!"

As we'll see later, the term *information* is also often used equivocally across contexts that may or may not involve selves. In physics and computer science, the term basically means a reduction in possibilities. A coin that falls on the floor heads-up is counted as one binary bit of information even if the result means nothing to anyone. By this definition, a flip of a computer switch or the collapse of a wave function in quantum physics is called information. Within their specializations, scientists are rigorous about the measurement of what they happen to call information.

Still, when selves are involved, the conventional meaning of information implies more than merely a reduction in possibilities. Information is useful for selves given their aims. So even in molecular biology, equivocation between the two senses of the term *information* can be problematic. This potential for equivocation can lead to confusion about whether minds are computers or computers are selves, and even whether the whole universe is a computer/self.

We have long been trapped, thinking of selves and aims as either invisible ghosts or mere cause-and-effect mechanisms. Equivocation enables us to pretend that we aren't trapped by these two overly simplified alternatives.

BRIDGELESS

We lack an explanatory bridge over a gap in our understanding, but the gap is not between two wholly independent realms. Selves and nonselves, aims and aimlessness are all natural phenomena. The bridge we lack connects two domains of one natural physical world.

Nonscientific thinkers supply all sorts of bridges that span that imagined gap, with supernatural ghosts breathing life into natural mechanics. Since science focuses exclusively on the natural, bridges from the supernatural to the natural are irrelevant here.

Not having a scientific solution to the mystery leaves the life sciences ungrounded. We, the scientifically inclined, urge people to embrace science and evolution as a bulwark against ungrounded supernatural thinking. In defense against religious fundamentalism, we argue that evolution provides an adequate explanation. To counter those who argue that the complexity of life required a supernatural creator-self, we argue that natural selection is explanation enough, sufficient to eliminate any need for supernatural selves and aims.

Still, when confronted with the mystery of purpose, we admit evolution does not create selves and aims; it just hones them. Natural selection is just our name for passive constraint imposed by circumstances biasing reproductive success. Evolutionary theory is therefore an incomplete alternative to supernatural creation myths.

Since evolution doesn't explain how selves and aims emerged in the first place, evolutionary theory hovers suspended a few inches above the physical grounding we claim for it. It dangles free, waiting for a scientific explanation for the origin of the selves and aims that evolution can then hone. We get by with a promissory note: someday scientists will explain how life emerged.

Only through a scientific solution to the mystery of purpose can we integrate the physical and life sciences and provide a grounded, comprehensive scientific model for how it is that we fit within the cosmic scheme of things.

It is notable that the "theory of everything" sought by contemporary physicists does not address the mystery of purpose. So it's not a theory of "everything" and, importantly, not a theory of what matters most to us.

As Nobel Prize–winning chemist Ilya Prigogine and philosopher Isabelle Stengers put it, "We must understand our world in such a way that it will not be absurd to claim that it has produced us."[1]

A solution to the mystery of purpose would yield far more than confidence that it's not absurd that we are here. Discovering what purposes, selves, and aims really are and how they work is a more grounded basis for putting our personal lives in perspective, for prioritizing effectively, for making better decisions and feeling better about the decisions we make. Under the influence of Deacon's solution, I have written over thirteen hundred blog articles about its everyday implications.

WEIGHTLESS YET WEIGHTY

You know you have aims and the moon doesn't. But have you ever seen an aim? You've seen the consequences of aims, but never the aim itself. An aim has no mass, volume, or charge. It's neither a material object nor a physical force. You know that your aims' existence depends upon materials, for example, neurons and neurochemicals. But the aim isn't these materials.

The self's existence also depends upon materials—those that compose the body. But the self isn't the body. At death, the self is gone but the body remains. A dead body weighs exactly as much as a live body.

Selves and aims are in fact weightless and yet of weighty consequence, and not just of importance to us selves but changing what actually occurs in the physical universe. No matter where you are, you're surrounded by countless physical objects that would be inexplicable without reference to the selves and aims that generated them. At home, at the office, on the busiest city street, or in the wildest woods, even in a seemingly barren desert, the fingerprint of selfhood is everywhere, molecules organized without defiance of any physical laws, yet inadequately explained by those laws alone.

Selves and aims are not the only weighty, yet weightless, nonthing things we employ to explain everyday phenomena. What, for example, is information? As mentioned earlier the term has taken on two meanings that allow a critical distinction to be ignored. Although there has been a very successful technical field called information theory, it has little to do with information as it is conventionally understood. The technical meaning of *information* is more accurately described as *signal theory* since it only focuses on characteristics of the signal or sign medium and ignores any meaning it conveys. Equivocating between the technical and the conventional use of the term *information* makes it possible to treat information as though it were a material thing or physical force.

Even with our best scientific instruments, we can't locate the meaning being conveyed. We can only see and measure the physical matter that "carries" it. Like the self, which is easily mistaken for the material body, information is easily mistaken for ink printed on paper, sound waves passing through air, photons emitted by computer screens, or electromagnetic waves flitting through space. These materials and material forces aren't the information. Whether carrying information or not, these physical or energetic media have the same physical qualities. Paper and ink weigh the same regardless of whether they convey information. Yet, we all talk of information as though we know what it is. We don't. We know what conveys it, and the consequences of its existence, but not what it is.

Whatever information is, it is only that to selves, given their aims. Dust mites employ information; molecules do not. Nothing is information *for* the moon, a galaxy, a quark, a pebble, or the whole universe. It's not just that we can't imagine a pebble interpreting information. It's that, in the absence of selves and their aims, a pebble changes only in accord with physical forces.

A KEY TO SOLVING OTHER BIG MYSTERIES

What's true of information is true of the many self- and aims-related means we employ to explain things. What, for example, is persuasion, sensation, feeling, yearning, value, significance, will, care, and even love?

One thing we know is that they're nothing to nonselves. Nothing persuades inanimate objects. They don't sense, feel, or value anything. Nothing is significant to them. They have no will or cares. They don't love or want anything.

We can't reference any of these factors in our explanations of purposeful behavior without also implying the selves who have them. In our guts we recognize the purposeful consequences they produce, but they are not the physical forces or the molecules that comprise them. Molecules don't care and care can't be reduced to molecules.

A solution to the mystery of purpose would ground our scientific understanding of all purposeful behavior, all of which is currently treated as ghostlike supernatural phenomena, as material mechanics, or, by equivocation, as both or neither, whenever it suits us.

A solution to the mystery would yield a scientific alternative to supernaturalism and explain the most significant transition at the origin of life. It would explain how evolution starts. And it would ground the debate on many of our most troublesome philosophical questions.

To take one example, do we have free will? The debate has remained unsettled for millennia—understandably, given the absence of an adequate scientific explanation of selves and their aims. Free or not, *will* emerges as a self and its aims. Inanimate matter and artifacts don't *will* anything. Selves obviously do.

Without a sound scientific explanation for how aims or wills emerge, arguments for the existence of free will remain scientifically ungrounded, while arguments against free will are based on the equally groundless assumption that all behavior is unwilled chemistry.

We have long underestimated the importance of having a science of selves, one that is more realistic than the assumption that we are ghosts in machines or just machines. Scientists acknowledge that the origin of life remains a mystery, but one among many treated largely as a molecular cause-and-effect problem that will be solved once we discover the right chemical interactions. We rarely hear big thinkers say that until we have a solution to the mystery of purpose, we're stuck in a holding pattern, circling above the big questions.

SUPERNATURAL GHOSTS, ELIMINATIVE MACHINES

Supernaturalism, the belief that everything matters to supernatural selves, is now resurgent in some parts of the world, perhaps in part because the scientific alternative seems to suggest that we are just complicated chemistry, mere matter in motion.

Lacking a solution to the mystery of purpose, some scientists treat all behavior as cause-and-effect events. By this reckoning, you are not a self, and you have no aims. Francis Crick, codiscoverer of the structure of DNA, said, "You, your joys and your sorrows, your memories and your ambitions, your sense of personal identity and free will, are in fact no more than the behavior of a vast assembly of nerve cells and their associated molecules. . . . This is in head-on contradiction to the religious beliefs of millions of human beings alive today."[2]

Believing, as Crick suggests, that selves and aims can be eliminated from scientific explanations, is called *eliminativism*. There are explicit advocates for eliminativism who have at times put it even more bluntly. For example, James Watson, Crick's partner in the discovery of DNA's structure, argued that "the essence of life is complicated chemistry and nothing more."[3]

If taken literally, Watson and Crick are right. Selves are indeed products of chemistry, but chemistry of a special sort, and not because they involve unprecedented chemicals. The atoms in our bodies follow the same chemical laws as the atoms in inanimate objects. An atom of carbon extracted from a living body would be indistinguishable from an atom of carbon extracted from a lump of coal.

Still, selves are not just any chemistry. They are *highly constrained* chemistry, chemistry limited to the self-regenerative interactions that keep us alive. To be sure, evolution is one source of this constraint, but since it doesn't solve the origin of life, it can't be the only constraint. The mystery of purpose is how general chemistry becomes self-directed, self-regenerative chemistry, the highly constrained chemistry of selves. But this mystery is sidestepped if one takes from the Watson and Crick quotations a more sweeping eliminative conclusion, that is, that selves are indistinguishable from chemistry.

Many eliminative theories focus primarily on explaining away human characteristics like beliefs and sensations as reducible to cause-and-effect events, without addressing the mystery of purpose, writ large. They assume that simpler organisms are just cause-and-effect chemistry evolved by the aimless process of natural selection. Neurophilosophers Patricia and Paul Churchland take this human-centric eliminative approach. For example, Patricia Churchland argues that "brains are not magical; they are causal machines."[4]

Eliminativism argues for the eventual elimination of means-to-ends explanations altogether. As science progresses, they argue, means-to-ends accounts will all be replaced by cause-and-effect explanations. By implication, then, nothing should really matter to any self, since selves are no different from any other chemical phenomena.

BAD PR AND BLINKERED SCIENCE

Obviously, there are big differences between inanimate chemistry and living selves. We admit it every time we refer to biological function (*function for selves, given their aims*), information (*significant for selves, given their aims*), or value (*good or bad for selves, given their aims*). Nothing is ever functional, significant, or adaptive for sodium chloride, snowflakes, mountains, fried chicken, or even computers.

Eliminativism is not just bad public relations for science; it's blinkered science. By claiming that selves and aims are illusory, scientists appear to imply that we don't really exist and that nothing has any real value. This is a serious blind spot if science is ever going to achieve full integration across its disciplines. But it also can block research into the mystery that is most relevant to us: the nature of what it means to be a self.

In summary, we have briefly touched upon three proposed solutions to the mystery of purpose that each fall short for science:

Supernaturalism: The assumption that there is a realm beyond the reach of the natural sciences that somehow breathes purpose into natural matter. There's a supernatural ghost in the machine, a higher power, soul, or spirit that animates matter.

Eliminativism: The assumption that there is no purpose anywhere. There is no ghost, just the cause-and-effect machinery of chemistry.

Equivocation: Using terms ambiguously, sometimes as though our aims are ghostlike purposes and sometimes as though they are mere mechanism.

If these approaches don't solve the mystery, what might? Deacon provides a plausible scientific alternative that demonstrates just how it is that we are neither ghosts nor machines.

LIKED VS. LIKELY

Life is meaningful or valuable to each of us. That's apparent in the way we aim to keep it going. A basic meaning of life for each of us, then, is not to end it. It's certainly not the only possible meaning or even always the highest priority. Some people willingly give their lives for other selves and aims, real or imagined.

And what happens to us when we die? A grounded speculation can be had only when we know what a self turns out to be, for how can we know what stops if we don't know what starts with life?

We may not want to know what happens when we die. It would be unrealistic to assume that we want to be that realistic. Visions of a supernatural afterlife have always served a variety of human aims. Supernaturalism isn't going to disappear once we solve the mystery of purpose.

Still, some of us aim to spend some portion of our lifetimes trying to discover what is going on here, regardless of whether it satisfies our personal aims. Aristotle's maxim "All men naturally desire knowledge" doesn't reflect our conflicting desire for hope and realism and the selectivity that results from the conflict, for example, between wanting and not wanting to know disappointing features of reality.[5] We quest scientifically for likely stories, but in our hearts, all of us also quest for likable or useful stories, which sometimes diverge from the likeliest stories. Truth doesn't always set us free.

In an apparently aimless universe, we humans have a rare opportunity. We are the only creatures we know that have the capacity to interpret carefully yet broadly and therefore to struggle scientifically in pursuit of an accurate account of the universe.

It's likely there's other life in the universe, but all evidence points to it proceeding elsewhere the way it started here, not with creatures with a human capacity to model reality through abstract language and thought, but with simple organisms. Though we have sought for decades to communicate with intelligent life elsewhere, we haven't succeeded yet, and some speculate that this is because language would be a rare, late-evolving trait, as it was here.

One great use of our short, languaged lives, an aim for those of us who can stomach and afford it, is to spend a little time on the many mysteries, chief among them who we are and how we emerged. So that's our focus here. We aren't born with packing slips that detail our contents. We're each born like amnesiacs with no idea who we are and how we got here. Taking notes on what we are is a fine aim.

3

DEACON'S SOLUTION IN BRIEF

AN UNROCKLIKE PERSISTENCE

Selves have persisted continuously on Earth for 3.8 billion years. If bodies were static objects like rocks, such persistence wouldn't be remarkable. But they aren't. Selves aren't inert, unchanging material structures. Our bodies are *dynamic processes.* The term *dynamics* can mean many things. Here I'll use it to mean large populations of interacting molecules that haven't settled into a static, stable, or resting state.

Selves are not even made of the same material from day to day. Our bodies are stirred by dynamic throughput of matter and energy. Beer drinkers joke about renting beer since it stays in them so briefly. Our bodies borrow everything: oxygen, water, protein, carbohydrates, and fats. Our "lease on life" entails "leasing" all of the energy and resources involved in our existence.

Though constantly stirred by energy and materials passing through, a living body exists intact and unified. This is what makes us selves despite all of this coming and going. Selves somehow have the capacity to regenerate their own selfhood throughout an individual life and, before dying, to pass their selfhood on to offspring.

Self-regeneration is how selves have persisted for so long, not as static objects at rest, but as dynamic paths of matter and energy that somehow stay the same despite material transience, and in the face of a relentless challenge: the tendency for everything to degenerate. A rock degenerates slowly because it's hard, solid, and stable. Without the self's regenerative capacity, our soft, unstable, dynamic bodies would degenerate rapidly, as occurs in decomposition at death.

At death, the self is gone and with it the capacity for self-regeneration. The capacity for self-regeneration is what it means to be a living self. Self-regeneration is the difference between life and death.

OUTPACING THE SECOND LAW

In our probabilistic universe, anything that can happen will happen, sooner or later. Unconstrained over vast lengths of time, however, the laws of physics and chemistry tend toward whatever results are most probable. And what is most probable is degeneration, not sustained self-regenerative organization.

Except for the special physical condition we call life, degeneration is the nearly inevitable result of physical change. Sorted things become unsorted, segregated things become desegregated, organized things become disorganized, concentrations become diluted and dissipate.

This tendency toward disorder was first discovered with respect to heat. Concentrated heat dissipates into its surroundings, a tendency called the *second law of thermodynamics.* Put hot and cold water together, and even with no stirring, the water becomes warm. The faster molecules in the hot region and the slower molecules in the cold region interact, becoming an irregular or mixed distribution of faster and slower molecules. Depending on the relative quantities of hot and cold water, and assuming no external heat is applied, the water sooner or later becomes uniformly tepid, not because all of the molecules are moving at the same velocity but because the speeds are randomly distributed.

Why? Because of all the possible arrangements of those water molecules, the vast majority are disorganized, mixed up—in other words, irregular. With every possible arrangement equally likely, irregular ones are what we get. It's like throwing a billion pennies into the air, one concentration regularized with tails up and the other concentration regularized with heads up. They could all land in some regular pattern of heads and tails, but that is certainly unlikely. It's far more likely that they'll land in an irregular pattern simply because there are vastly more irregular than regular patterns possible. Given the second law, dynamics fall toward irregularity as readily as balls roll downhill.

The second law isn't a force. Nothing and no one enforces it. The second law is the consequence of the absence of constraint, the lack of a force or enforcement, nothing and no one constraining what happens. It is simply a description of what tends to happen when there is no force, constraint, or other influence to prevent irregularity.

The second law tendency applies to all kinds of energetic processes. Pressure equalizes, batteries discharge, radioactivity decays; beyond energy, it applies to everything that can change. Even rocks degrade. Their molecules dissipate slowly as crystalline bonds break.

Here I'll generalize the second law beyond its energetics applications. I'll refer to the general tendency toward irregularity or mixed-upness as the second law. (The first law is about what doesn't change, that is, the *quantity* of matter or energy involved.)

Scientists agree that the second law is as fundamental as physical laws get. The mystery of purpose is that second law irregularity is the opposite of what we see in selves. Your body is no random molecular coin toss. Tossed by continuous dynamic throughput, your body somehow remains highly regularized.

Self-regeneration outpaces the second law such that selves not only survive but also reproduce, proliferating populations of selves from a single ancestor, here, and quite likely in innumerable places in the universe, wherever selves have emerged.

When selves lose their capacity to self-regenerate they die, though not necessarily before passing on to offspring their self-regenerative capacity to work against, and outpace, the second law. Self-regeneration is a capacity to locally defy the second law, temporarily in each individual self, and sustainably over the history of life. Bodies have evolved into Darwin's "endless forms most beautiful and most wonderful" over that history, with self-regeneration sustained uninterruptedly all the while.

How, then, in a universe in which the vast majority of possibilities for change tend toward maximum irregularity do we ever get the highly regularized states evident in selves?

EMERGENT REGULARIZATION

Under certain conditions, the tendency for dynamics to become irregular reverses. For example, when a stream flows around an obstacle, it briefly forms turbulent (irregular) currents and then settles into a whirlpool path, with water flowing in a regularized spiral pattern that's not imposed by an external constraint. This spontaneously forming orderliness catches our attention. It's as though Galileo started a ball rolling down a ramp and, without any change to the ramp, the ball suddenly reversed direction halfway through his experiment.

What causes the whirlpool to form, if it is not being stirred in a circle? What exactly imposes the constraint on how the water flows such that, rather than staying turbulent and irregular as we might expect given the second law, it instead becomes regularized into a spiral current?

The spiral-forming constraint is not imposed. There's nothing pushing, pulling, or stirring the turbulent currents, molding them into a spiral. The whirlpool is *emergent*, meaning that it *arises from the interactions taking place throughout the dynamics.* In the case of a whirlpool, it arises from the interaction between the water molecules.

Turbulent water currents get in one another's way, generating impasses, congestion, or gridlock, which slows and stops some currents. Conflicting currents that were initially common become progressively less common while the spiral current remains, having become relatively more likely because they do not get in one another's way—the path of relatively less resistance for water throughput.

The whirlpool's spiral current may seem to be caused by some thing or force added or imposed from the outside upon the water flow, but it isn't. It's the elimination of alternative paths—likely paths becoming unlikely—not because new constraints were imposed, but because new constraints *emerged* through dynamic interaction.

That's how I'll use the term *emergence* throughout this book. Constraints can be imposed, or they can emerge from interactions occurring throughout dynamics that progressively reduce the variety of likely currents.

Emergent constraint is decidedly different from imposed constraint. With the whirlpool, the imposed constraints include the running water, whatever channel the water runs down, and the obstacle in the channel. These imposed constraints are necessary but insufficient conditions for explaining the whirlpool's regularized spiral flow, a regularity that emerges from the congestion throughout the turbulence and the ways that the congestion constrains the paths that the flowing water can take.

The whirlpool is an example of *emergent regularization*, a regularity—the spiral flow—that remains after other paths have impeded one another. Emergent regularization will be important for solving the mystery of purpose.

But emergent regularization is only one of two kinds of emergent constraint. The sort of emergent constraint exhibited by a whirlpool is commonly known as *self-organization*, a term coined by systems scientist W. Ross Ashby, who argued, "a self-organizing system spontaneously reduces . . . the number of its potential states."[1] Whirlpools are but one example of self-organization. We'll visit a few more shortly.

Unfortunately, self-organization is not a great name for the process. There's no self that organizes the water into a whirlpool, nothing imposes the organization, and "organization" is an ambiguous term. If it means arriving at a perfectly organized state, that's not what happens. The current is only more or less regularized. And organization can refer to whatever configuration dynamics might take, for example, an irregular organization. So I will refer to self-organization as *emergent regularization* instead.

Many researchers assume that emergent regularization (self-organization) will prove sufficient to explain the origins of life.[2] Get enough regularization going and it either becomes a real self or, if not, at least dynamics that are evolvable by natural selection.

But life isn't just regularization. With selves, something else emerges, the capacity for *self-regeneration*. With the whirlpool, water flows in a constrained spiral pattern, but there's nothing about that spiral that regenerates the spiral if perturbed. If you change the flow or reposition the obstruction, the whirlpool does nothing to resist these modifications. The whirlpool does no self-directed work.

We need therefore to distinguish *emergent regularization* from *emergent self-regeneration*, a second kind of emergent constraint. Regularization and self-regeneration are both *emergent*, meaning that they're constraints that aren't imposed, but instead arise throughout dynamic interaction. Yet what emerges from each is different. With emergent regularization, all that emerges is an ephemeral regularity like the whirlpool's spiral. With emergent self-regeneration, a self emerges. Just how this happens is what must be explained if we are to solve the mystery of purpose.

EMERGENT SELF-REGENERATION

Emergent self-regeneration is the emergence of constraints that channel energy into work that regenerates these selfsame constraints. That's what selves do. We regenerate ourselves by constraining, indeed, by *aiming* our work.

In physics, work is due to the constrained release of energy—as we shall see later, constrained second law degeneration. Selves constrain the release of energy into work aimed at regenerating their ability to constrain the release of energy into work aimed at regenerating that ability—a circular dynamic that's been continuously maintained for billions of years since the first selves emerged on earth.

Self-regeneration is thus our first and foremost aim. It's our first means and end, and it's circular: the means by which we can regenerate our means, the end or purpose being the ability to continue to pursue our ends.

In contrast, a whirlpool is actually degenerative. It constrains water flow into work that creates a spiral that is the water's path of least resistance. This can be seen in the case of the whirlpool that tends to form as water exits a bathtub. Water drains more rapidly with a whirlpool than with turbulence at the drain. So a living self, modeled as emergent regularization, like a whirlpool, would be a short-lived, self-eliminating self.

The whirlpool is a process of elimination that eliminates itself by more efficiently converting energy into work that efficiently degrades the necessary conditions for its existence. And whirlpool offspring? Of course, there are none. As we'll see later, all examples of emergent regularization are degenerative like the whirlpool.

Emergent self-regeneration is different from emergent regularization, and not just because it generates offspring that carry forward self-regeneration over generations. Emergent self-regeneration yields three capacities that emergent regularization doesn't yield:

Self-repair: The ability to regenerate regularizations faster than they would otherwise degenerate, given the second law. The second law never stops degenerating things. For us soft selves to exist, we have to outpace second law degeneration with self-repair.

Self-protection: The ability to resist or withstand the second law tendency toward degeneration, for example, by generating protective tissue, cell walls, skin, exoskeleton, epidermis, or bark. To some extent, selves can endure even in conditions where there is no supportive environment. In contrast, a whirlpool disappears as soon as water flow ceases.

Self-reproduction: The ability to pass on the capacity for self-regeneration to multiple varied offspring.

THE CIRCULAR SELF

We all recognize that there's something circular about being alive—the regeneration of the power to regenerate. We think of it mostly in generational terms—offspring becoming parents that produce offspring that become parents.

We don't tend to notice that regenerative circularity—regenerating our ability to regenerate—must operate in real time throughout an individual self's life. We have to regenerate in real time because we're mostly soft tissue. We don't tend to think of self-repair or healing as an essential feature of all life. We might think it's essential to human life in that when we lose the ability to heal, we die. But we tend to overlook the fact that self-repair is necessary for all selves, even in the absence of injury and even for the selves that emerged prior to evolution by natural selection.

Self-repair is even a higher priority than self-reproduction. A self can't reproduce if it doesn't persist for long, and it can't persist without self-repair because it is constantly at the mercy of the second law. Self-repair presents a serious challenge to the popular hypothesis that life begins when individual RNA molecules start replicating. Individual molecules are passive. They don't repair themselves.

There's a fundamental conflict between self-regeneration's three capacities. Self-repair and self-reproduction require a throughput of energy and resources. Selves need to be open in order to maintain access to these. But selves can't afford complete openness or the second law wins, and they just degenerate. That's why selves need self-protection too, a limit on the energy and resources with which they interact. They need, therefore, what I'll call *selective interaction*—openness to the right, and not the wrong, interactions for maintaining their capacity for self-regeneration overall.

It's sometimes said that you should keep an open mind, but not so open that your brains fall out. Something similar can be said about selves. Selves need to keep an open body, but not too open—open to interactions conducive to self-repair and self-reproduction, but closed to interactions such that self-protection is maintained.

Evolved cells achieve selective interaction by means of highly evolved specialized protein complexes embedded in cell membranes that allow for selective permeability. The selective transport of materials in and out gets more complex in multicellular organisms. Plants have evolved specialized pores, and animals have evolved orifices and other partially protective tissues that enable selective interaction with our environments.

But the selves that emerged prior to evolution had no opportunity to evolve these complicated means for selective interaction and selective interaction is highly unlikely to have emerged by chance chemistry, given the precision it requires. We can't simply assume that selves emerged with a semipermeable cell membrane that allowed just the right interactions with their environments.

AN EMERGENT CONSTRAINT ON EMERGENT CONSTRAINTS

Self-regeneration, with its three capacities—*self-repair, self-protection,* and *self-reproduction*—and its overarching challenge, *selective interaction*, specifies what selves must achieve, but not how it is achieved.

To summarize our exploration so far:

In a universe governed by the second law, irregularity is the most probable result.

Constraints explain how regularities can occur. *Imposed constraint* is the most obvious source of constraint. It's easy, for example, to understand that if we impose a solid channel on water flow, it will become regularized rather than flowing every which way.

But constraints can also *emerge*, meaning that, throughout dynamic interaction, some possible paths become less likely, thereby leaving remaining paths more likely, even without new constraints being imposed. Thus, there are two sources of constraint: one is imposed from without and the other emerges *throughout* dynamic interaction.

Emergent regularization (that is, self-organization) has been a hot research topic in recent years. Some researchers argue or assume that it can explain the origin of life. But by itself, it can't. Emergent regularization is ultimately degenerative, the opposite of self-regeneration.

Since regularization is inherently degenerative, it doesn't explain self-regeneration. Still, the fact that regularization need not be imposed but can emerge from dynamic interaction suggests a solution to the mystery of purpose that has been long overlooked. Like the regularization we find in whirlpools, perhaps self-regeneration is the product of a second kind of emergent constraint.

CONSTRAINTS ACCUMULATE

Emergent regularization results when countertendencies impede one another. We see this in the whirlpool's turbulent currents getting in one another's way. Is it possible that self-regeneration emerges from countertendencies between underlying emergent regularizing processes?

That's what Deacon imagines as the origin of the first self.

He pictures two emergent regularizing tendencies working for and against each other such that each prevents the other's tendency toward degeneration. He demonstrates how two emergent regularization dynamics (known as autocatalysis and self-assembly) can be synergistically coupled such that they constrain each other's tendency to degenerate.

The result is a higher-level emergent constraint, an emergent constraint that further constrains the two underlying emergent regularization dynamics. This higher-level emergent constraint results in a tendency to continually regenerate itself by eliminating or constraining the lower-level emergent regularization tendencies toward degeneration. Deacon calls his model for this emergent self an *autogen*, in other words, a "self-generator."

Later, I'll show how the autogen achieves all of the capacities required for self-regeneration: self-repair, self-protection, self-reproduction, and selective interaction. I'll also present Deacon's speculations about the autogen's first evolvable traits, including the incorporation of information-bearing molecules.

The autogen is a thought experiment intended as a proof of principle that true selves with real aims can emerge within basic physics and chemistry. Unlike many thought experiments, this one is empirically testable.

As the simplest instance of spontaneous emergent self-regeneration, the autogen has true purpose—not purpose for the universe or for some other entity that wills it, not purpose that we as outside observers imagine by equivocation, and not purpose it has chosen for itself, but rather emergent purpose, purpose that emerges by chance chemistry.

The autogen, a spontaneous chemical process, nonetheless engages in self-directed work. It is an identifiable locus of self-control or agency, the means by which it can self-regenerate. If you're eager, you can skip straight to part 5 to see how autogens achieve self-regeneration. Also, a walk through the appendix at any time can provide orientation to the arc of Deacon's argument.

With this brisk introduction out of the way, we'll now expand toward a more careful explanation of Deacon's solution. We must move carefully because the path to solving the mystery of purpose is slippery, a bit like walking along a ridge trail with plenty of places where one can slip off in either of two directions: down one slippery slope into supernaturalism or equivocation (assumed ghosts in chemical machines) or down the other slippery slope into eliminativism (no ghosts, just chemical cause-and-effect mechanisms).

II

FRAMING THE MYSTERY

4

TWO SOURCES OF CHANGE

THE CAUSE-AND-EFFECT TOOLKIT

In explaining change, people employ two toolkits: a cause-and-effect toolkit for cause-and-effect events, a means-to-ends toolkit for means-to-ends behavior. With the former, every effect is traceable to prior material causes. The wind caused the window to rattle; the rain caused the dirt to become mud; the hammer fell, causing a dent in the floor.

In all cases, cause happens first; effect happens after. Given particular physical laws, because *Material Thing X* moved first (prior cause), *Material Thing Y* moved next (subsequent effect), or because there was a change in the interaction between *Material Things X and Y* (prior cause) both things changed (subsequent effects).

Thus we can explain the cause-and-effect interaction between two billiard balls by saying that the cue ball moves (prior cause), striking the eight ball, causing it to move (subsequent effect) in accordance with Newton's laws of motion.

With the cause-and-effect toolkit, we can string together sequences of change, explaining domino effects and Rube Goldberg–like chains of events. We can funnel in, explaining how multiple, prior material causes converge to produce an effect, or we can fan out, explaining how a prior, material cause produces multiple material effects. With all of these variations, the theme is material objects interacting, pushing or pulling on each other with highly predictable results.

Though we don't detail every aspect of a cause-and-effect interaction, we sense that we could. For example, two billiard balls are actually two constellations of molecules tightly bonded into spheres. On impact, both constellations are deformed. In explaining the interaction between the balls, we might ignore the deformity that results. Still, we're confident that cause and effect can explain the molecular- and even atomic-level interactions within the billiard balls, providing a fine-grained account of the pushes and pulls between material objects at any scale.

Only a few centuries old, our cause-and-effect toolkit readily explains events at the scales where we also find means-to-ends behavior. Newton's laws, electromagnetism, gravity, thermodynamics, chemistry, and statistical mechanics—these and other physical science tools make up the toolkit of reliable, as-good-as-timeless, universal laws for explaining and predicting cause-and-effect material or mechanistic change. Still, for means-to-ends behavior—behavior that matters to selves—we employ another toolkit, which, at present, is more like a magician's bag of tricks.

THE MEANS-TO-ENDS TOOLKIT

Cause and effect explains why the eight ball moved, but why, of all places, did it move toward the corner pocket? To explain this, we employ a means-to-ends toolkit and flip the explanatory sequence. We say that the pool player's *aim* to win the game in the future (subsequent end) caused her to aim the eight ball toward the corner pocket (prior means).

With cause-and-effect events, prior causes produce subsequent effects. Inversely, with means-to-ends behavior, subsequent ends produce prior means, and not nearly as predictably. The pool player can miss.

Of course, the future can't cause the past, so we rightly intuit that aims are *represented* somehow within selves. Thus, we can translate our means-to-ends, backward-in-time sequences into reliable, familiar, cause-and-effect, forward-in-time sequences.

We do this using special purpose-related terms from the means-to-ends toolkit, such as *wants*, *needs*, *ambitions*, *values*, *desires*, *intentions*, *appetites*, *goals*, *yearnings*, *aspirations*, *will*, *hopes*, *purposes*, and *preferences*. Thus, we might say that the pool player's desire (prior cause) caused the eight ball to move toward the corner pocket (subsequent effect).

But what and where is a desire? We can locate the consequence of a desire, but we can't locate a material object that is the desire. We assume that desires or more generally aims are located within our living matter—in our heart, mind, or somehow "hardwired" into us. A heart is a material thing that we can point to, but where is the desire within it? We know that an aim is associated with certain hormonal and neurochemical secretions, but we also recognize that an aim isn't itself a chemical secretion.

Our brains are material objects, so maybe our aims are in our minds, which are in our brains. But again, dissect the gray matter of brains, and you won't find a mind or an aim. Aims have no momentum, mass, charge, or energy to initiate material change. To manage the immaterial, mysterious nature of aims, we're compelled to treat them as though they were nonmaterial, ghostlike causes of material effects.

BLURRING THE TOOLKITS

In our everyday lives, we don't need to know what aims are. Pretending that aims operate like material causes works well enough. In explaining anything, we employ causal combinations from the two explanatory toolkits, always with material cause and effect as the model: prior change in one object causes a subsequent change in another object. For example, aims move the cue ball, which moves the eight ball toward the corner pocket. It's all material cause and effect to us.

We assume we can explain behavior by modeling it as a cause-and-effect sequence of any length using any combination of material objects and aims. For example: "My desire (aim) for cake (material) caused me to eat it (material). Eating it caused me to get fat (material), which caused me to want (aim) to exercise (material), which caused me to feel encouraged (aim), which caused me to wish (aim) to buy exercise equipment (material)."

Our everyday tolerance for mixing and matching is evident in the blurring metaphors we use to describe how aims cause effects. Treating aims like physical forces, we say that our wants and needs *pushed or pulled* us to act, our goals *propel* us, and our desires *suck us* in. We say that ideas have *impact* or leave an *impression.* We speak of attraction as *good chemistry,* love as *good energy,* and influence as *power.* In this way, we operate on the assumption that all change, whether consequential or not, is a product of pushes and pulls, the cause-and-effect interaction between objects, some material and others immaterial.

We also do the opposite, treating cause-and-effect events as though they were means-to-ends, aimed, or purposeful behaviors. The river seeks the sea; the mountains reach for the sky. We blur even in the sciences, for example, when saying that molecules have *affinities*, as when we describe lipid molecules as having *hydrophilic* (water-loving) and *hydrophobic* (water-fearing) ends.

Thus we animate, treating cause and effect as means to ends, but we also *deanimate*, treating aims as though they were material objects and forces that act by cause and effect on other materials.

For science, this mix-and-match equivocation of causes and aims is deeply problematic. If our aims are just causes in cause-and-effect sequences, they sure are weird ones. For one thing, ends don't yield reliable outcomes the way material causes do. The eight ball always moves the same way when struck the same way by the cue ball, but a pool player's aims don't cause their effects anywhere near as reliably. As the pool player's mother used to say, an aim doesn't cause you to get what you want. Furthermore, while cause-and-effect explanations apply to all physical matter, our means-to-ends explanations only apply to selves.

5

SELVES

SELF REDEFINED BROADLY

Your selfhood is constant throughout your life despite the matter and energy coming and going through you, the changes to your character, the learning, growing, and aging, the gain and loss of faculties and even body parts over your lifetime. You are the same self from conception to death.

All humans are selves, but are all selves human? By conventional definition, mostly yes. You might call an ape, chimp, horse, cat, or dog a self, but perhaps not a chicken, spider, worm, or sponge, and likely not a tree, fungi, or bacterium. People draw the line between selves and nonselves with humans squarely on the "self" side, though with a little ambiguity about who or what else counts as a self.

Because humans have self-awareness, they must be selves. Some argue that self-awareness is all there is to selfhood. For example, in some philosophical and spiritual (for example, Buddhist) circles, selves aren't considered real at all. There's just self-awareness, the false impression that one is a self.

Here, I'll assume that selves are real and that selfhood encompasses far more than self-awareness, consciousness, ego, or any other psychological characteristics. I'll regard all living beings—and most important for solving the mystery of purpose, the very first living beings—as selves.

Selfhood is what's constant throughout any life despite the matter and energy coming and going throughout a self's body, despite changes to character, learning, growing, and aging, or the gain and loss of faculties and even body parts over a lifetime.

Selfhood is also the constant throughout the natural history of life on earth from the first, preevolutionary self forward. Selfhood is what has been passed on from generation to generation continuously for 3.8 billion years.

Selves go by other names. People call them individuals, beings, organisms, creatures, or agents. Any term we choose has its connotations, its potential to encompass too much or too little. *Individual* can encompass individual inanimate objects too, for example, individual rocks. *Being* is broad and vague, potentially connoting anything that exists. *Organism* implies the material body more than the self. *Creature* isn't very descriptive and can suggest something created, even a machine. *Agents* have agency or aims, but we also speak of chemicals and computers as agents.

HUMAN VS. NONHUMAN SELVES

To face the mystery squarely, I'll use the term *self*, despite its psychological connotations. I broaden the class of all selves in order to jog us out of an intellectual habit that has ungrounded philosophy and science for millennia, the assumption that the mystery of purpose applies chiefly or exclusively to human selves. It doesn't. Purpose only applies to selves, but selves include all living beings.

We humans are indeed radically different from other organisms. Terrence Deacon, whose theory this book presents, is well aware of the difference between humans and other selves. His first book, *The Symbolic Species*, explored and explained the origin of human language, a necessary condition for the distinctive features of human selfhood. According to Deacon, "Biologically, we are just another ape. Mentally, we are a new phylum of organisms."[1]

Still, Deacon recognizes that purposeful behavior doesn't emerge with human consciousness but must have emerged at the very origin of life. The mystery of purpose must be solved with an explanation of selves, broadly defined to encompass all of us, not just humans. To Deacon: "Self is, in all cases, the origin, target, and beneficiary of functional organization. Thus, there is good reason to believe that by first exploring self at its most basic level we may be able to discern some fundamental principles that will apply as we build our analysis upward toward the most complex phenomena of selfhood: human consciousness."[2]

By encompassing all organisms as selves, we gain a perspective undistorted by a focus on humans, a highly evolved and most unusual species of selves. Deacon argues that trying to explain purpose starting with humans is like trying to understand hair starting with porcupine quills. Quills are indeed hair but highly evolved and most atypical. Deacon suggests that by focusing on purpose, selves, and aims at their very origin we are likely to gain crucial insights helpful to any exploration of human nature because "Knowing how something originated often is the best clue to how it works."[3]

THE SELF-BODY PROBLEM

The influential Enlightenment philosopher René Descartes contributed mightily to our misguided tendency to treat the mystery of purpose as a problem that must be solved with an explanation of what distinguishes humans from nonhumans, not all selves from nonselves.

Descartes argued that all plants and animals are material mechanisms but that humans are different, rational in a way that other creatures are not—rationality being, by his account, the ability "to act in all the contingencies of life in the way in which our reason makes us act."[4]

To Descartes, we humans think, therefore we are. Plants and animals comprise one substance, but we humans comprise two. Other beings are machines. According to Descartes, we humans are machines but also the power to reason, a second, nonmaterial substance that acts upon us by cause and effect, making us act.

Descartes lent prominent and persistent credence to an approach called substance dualism, meaning two substances. He called one substance *res cogitans*, Latin for *thinking substance*. He called the other substance *res extensa*, Latin for *substance extended in three-dimensional space*—in other words, the material thing.

To Descartes, *res cogitans*—the mind—was not material. He argued, "The rational soul . . . could not be in any way extracted from the power of matter . . . but must . . . be expressly created,"[5] created, that is, by a rational, purposeful creator, a God, who aimed to endow us with rationality.

Descartes thus cornered philosophers with what became known as the mind-body problem—the challenge of figuring out how the two substances, the thinking and extended things, interact yielding conscious matter. Philosopher Gilbert Ryle coined the phrase "ghost in the machine" to expose the problems with Descartes's mind-body substance dualism. What is the ghost, and how does it get into the machine? Ryle said, "Such in outline is the official theory. I shall often speak of it, with deliberate abusiveness, as 'the dogma of the Ghost in the Machine.' I hope to prove that it is entirely false, and false not in detail but in principle."[6]

Biophilosopher Gregory Bateson and others argued that there is some mind in every organism, not just in those with the capacity to think.[7] To Bateson, mind emerged with life.

We might debate whether a bacterium exhibits mind, but we're all clear that organisms are alive, selves with aims as we define them here. Thus, the mind-body question becomes the self-body question. How does a material, dynamic mechanistic body end up also being a self with aims? One might try to resolve it by rejecting dualism and claiming that the body is the self.

"I AM THE BODY I HAVE"

Philosopher Daniel Dennett framed mystery of purpose this way: "Now there are selves. There was a time, thousands (or millions, or billions) of years ago when there were none—at least none on this planet. So there has to be—as a matter of logic—a true story to be told about how there came to be creatures with selves."[8]

By assuming that all creatures are selves, I'll go with billions of years—starting from the origin of life, not of human self-awareness. I'll assume Dennett's "creatures with selves" coemerged since there are no creatures that aren't also selves.

Dennett's peculiar phrase "creatures with selves" reflects his focus on how creatures evolved to possess self-awareness. If, instead, by *creatures* we mean bodies, then we're cornered appropriately with the self-body question with but one remaining escape route. Isn't the self just the body?

We express ambivalence about our answer to this question in the way we alternate between saying that we *are* our bodies and saying that we *have* our bodies. If we answer that we are our bodies, we are merely Ryle's machines.

When instead we say that we have our bodies, we acknowledge the mystery that must be solved, though not with mysterious ghosts. We acknowledge that the self continues to exist while the material that makes up the body changes day to day. We can lose and replace parts of our body, while our self remains intact. Most important, when we die, the self is gone, but the body is still there.

If we aren't our bodies but are, instead, the selves embodied by bodies, then what are we? That's the mystery. Again, selves are not material objects. Dead or alive, a body has the same material qualities. Like a ghost or soul then, the self has no mass or volume.

Aims are mysteriously immaterial too. What does it mean to have a desire, appetite, or purpose when we can't put a material finger on such things? Even if we don't believe in ghosts, we can't seem to avoid explaining our behavior without them, equivocating between treating them as material objects and treating them as phantoms. In conventional treatment, selves and aims are oxymorons: immaterial materials, nonsubstantive substances, nonthing thingies, and immaterial causes of material effects.

And we will need to put a finer point on the ghost and machine distinction because, in the current debate, we confuse two kinds of ghosts and two kinds of machines.

6

TWO GHOSTS, TWO MACHINES

SUPERNATURAL VS. EQUIVOCAL GHOSTS

We associate ghosts with the occult, meaning hidden or concealed, inaccessible to empirical investigation. As such ghosts are denizens of the supernatural, a realm beyond the natural, members of an extended family of selves that includes souls, spirits, angels, and gods.

There have long been supernatural solutions to the mystery of purpose, for example, that a supernatural God-self created our supernatural souls to serve His aims. By this account, the self-body problem is easily solved. Your body is a machine built to serve your supernatural soul's aims. Your supernatural soul was created by a supernatural God to serve His aims.

Today, a popular supernatural approach replaces God with a *higher power*, though never imagined as a mere voltage. It always has aims. Thus, though imagined as amorphous and omnipresent, trimmed of its body and white beard, the higher power is still unimaginable as anything but a self.

There's something like a second ghost story that even scientists often tolerate, though often unwittingly—ghosts by analogy. Every time we use a personal pronoun or an aim-related word like *need*, *want*, *drive*, or *desire* we implicate something ghostlike.

Those of us who reject supernatural explanations are confident that whatever selves and aims are, they're natural phenomena. We all employ the means-to-ends toolkit to explain things that occur in nature, even though we are unable to explain what's in the toolkit. We thus rely on what I'll call *equivocal ghosts*. Equivocal ghosts are natural, not supernatural, yet their nature is ambiguous and elusive. Equivocal ghosts are what selves and aims are at present in everyday, and even scientific, discourse.

In science, equivocal ghost-selves are called *homunculi*, meaning "little men." As philosopher Daniel Dennett remarks, "Wherever a theory relies on a formulation bearing the logical marks of intentionality, there a little man is concealed."[1]

EMERGENT HOMUNCULI

Eliminative approaches campaign to purge homunculi from science because they are ambiguous and unexplained and therefore not a scientific explanation for anything. The eliminative campaign has not been successful, nor can it be since the consequences of these equivocal ghosts are inescapably apparent and are not reducible to material cause-and-effect phenomena.

To solve the mystery of purpose scientifically requires that we reject all supernatural ghosts, but also all equivocal ghosts, the little men rife even in scientific discourse.

Still, in expunging all such unscientific ghosts, we are left with true selfhood to explain, the selves we know by the fruit of their aimed labor. Thus, ours is not to bury all homunculi but to explain the real ones, interpreting the currently equivocal ghost-selves in such a way that they become unequivocally explained natural phenomena.

Equivocal ghosts whose consequences we recognize so readily both through personal experience and within the life and behavioral sciences are what must be explained, not explained away. Until they are explained, we should admit that they're homunculi, equivocal ghosts that we know by their consequences alone.

An approach called *emergentism* admits that we don't know what selves and aims are yet and that it would be worth knowing. *Emergentism* seeks an explanation for key transitions from one kind of phenomena to another, chiefly from cause-and-effect dynamics to selves with aims.

Deacon is committed to emergentism, arguing that the burden of proof is indeed on science to explain the nature of selves and aims. Emergentism is the honest scientific alternative to eliminativism, which explains away selves and aims, to supernaturalism, which explains selves and aims as originating in a realm beyond and permanently inaccessible to science, and to equivocation, which sidesteps the mystery of purpose through ambiguous terminology. Emergentism puts the burden of explaining selves and aims squarely on scientists' shoulders, where it belongs.

FUNCTIONAL VS. NONFUNCTIONAL MACHINES

There are two kinds of ghosts—supernatural and equivocal—and loosely speaking, there are two kinds of machines, nonfunctional and functional. The machine in the phrase "ghost in the machine" refers to nonfunctional dynamics, dynamics again meaning the cause-and-effect interactions that occur throughout large quantities of material components that have not settled into a stable static state. There are dynamics that serve and don't serve selves and aims, as well as gradations in between the two kinds of dynamics.

Dust storms, galaxies, and exploding stars are but a few examples of nonfunctional dynamics, dynamics that occur regardless of there being any selves whose aims they serve. Computers, cars, and other machines are examples of dynamics that serve selves.

In between, there are dynamics that can incidentally serve selves. A rainstorm by itself is just inanimate, nonfunctional dynamics though it can come to function for selves. Then there are dynamics modified by selves to serve our aims. Whereas a river, by itself, is just nonfunctional dynamics, a river we divert to irrigate our crops becomes functional dynamics.

Machines or artifacts are dynamics engineered and constructed by selves exclusively for their uses. Unlike the rainstorm or diverted river, machines are highly improbable dynamics in the natural world. A car engine or computer is highly unlikely to appear by happenstance in the inanimate universe, and today's millions of nearly identical computers are surely not going to appear by happenstance.

It's also possible to equivocate on the difference between nonfunctional and functional dynamics with terms like *mechanism* or *mechanics*. *Mechanism* tends to imply function but need not. One could address the mechanisms by which stars or molecules form or rains fall. Mechanics can refer to functional engineering but also to functionless dynamics, as in Newtonian, statistical, or quantum mechanics.

Now consider the mechanisms of living bodies. Like machines, they are highly improbable dynamics. But unlike machines, they aren't engineered. Still, many if not most explanations for the existence of selves, even from scientific sources, employ the functional machine metaphor. For example, biologist John Maynard Smith gives voice to a popular, though equivocal, explanation when he states that "DNA contains information that has been programmed by natural selection."[2]

Programming is engineering that selves do with machines in order to serve engineers' and users' aims. But natural selection is no engineer. One can claim that referring to bodies as machines is an innocuous metaphor that makes functional dynamics easier to understand. Easier, yes—from everyday experience, we have strong intuitions about how engineers control dynamics to produce functional machines. But innocuous, no, at least if we

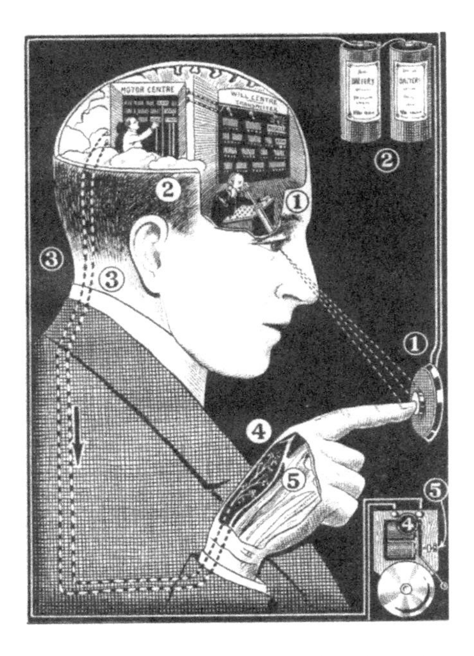

FIGURE 1 An intuitive treatment of selves as machinelike, though with a homunculus programming the machine.

are to avoid equivocation in our attempts to explain the machinelike functional intricacy of bodies that *aren't* engineered.

TELEONOMY

Many biologists, including Ernst Mayr, are ambivalent about treating purpose as a natural phenomenon. Mayr and others attempt to sidestep true natural teleology or purpose by accounting for purposelike behavior with the term *teleonomy*. The *-onomy* ending comes from *nomos*, meaning "law." Thus, teleonomy means the impression of purpose resulting solely from the laws of nature.

The term was coined by Colin Pittendrigh,[3] but given focus by Mayr, who restricted it to "systems operating on the basis of a program of coded information."[4] Mayr illustrated the distinction between teleology and teleonomy by arguing that it is teleological to say that "the Wood Thrush migrates in the fall in order to escape the inclemency of the weather," but teleonomical to say that "the Wood Thrush migrates in the fall and thereby escapes the inclemency of the weather."[5]

Teleonomy suggests that the behaviors that we mistakenly consider to reflect aims are reducible to cause-and-effect dynamics. But are aims fully eliminated in this account? If so, what is the wood thrush that benefits from escaping the inclemency by means of its programming? And what are programming and coding anyway? We know what it means to program computers. Selves do so to serve their aims. Who programs the wood thrush? If there's no one whose aims are served by programming the wood thrush, what justifies calling it programming, not cause and effect?

Philosopher David Hull found the distinction between teleology and teleonomy dubious and quipped:

> [J. B. S.] Haldane [like Mayr, another Neo-Darwinian luminary] can be found remarking, "Teleology is like a mistress to a biologist: he cannot live without her but he's unwilling to be seen with her in public." Today the mistress has become a lawfully wedded wife. Biologists no longer feel obligated to apologize for their use of teleological language; they flaunt it. The only concession which they make to its disreputable past is to rename it "teleonomy."[6]

It's time for biologists and the rest of us to dispense with such euphemistic, equivocal treatment of telos or purpose. But to do so we'll have to draw a clearer distinction between it and nonpurposeful phenomenon. For that, we'll next explore for-ness and about-ness, qualities that only apply when selves and aims, the sources and beneficiaries of true telos, are involved.

7

INTERPRETATION

FOR-NESS

Of value for, good for, bad for, functional for, useful for, significant for, information for—these only make sense when selves of some sort follow. Things are *for* selves and selves only. Nothing is of value or significance for a rock, an atom, a galaxy or the universe as a whole. It doesn't make sense to say that something is good for, or significant for, even the fastest supercomputer programmed with the best current approximation of artificial intelligence.

No matter how good artificial intelligence ever becomes there's no reason to start wondering whether anything is of value or significance for it until it somehow became a true self. What would that take? Most fundamentally the computer would need to have aims of its own and do self-directed work that is at risk of being thwarted. Were we to dismantle even the "smartest" computer we have, it wouldn't resist. It has no skin in its own game. Indeed, it has no game of its own.

Though it's obvious when we stop to think about it, it's worth noting that explaining for-ness is not a question about the function of an object but rather the beneficiary of that function. One could mistakenly argue that a hammer has for-ness since it's good for things. But that's not the issue. A hammer is good for things only for selves given their aims. Without selves who benefit, a hammer is not *for* anything.

ABOUT-NESS

Life scientists attend to fitness in a way that no physical scientist would. You'll never hear physical scientists talking about atoms, molecules or planets adapted to *fit* their circumstances.

What then is fitness? "Survival of the fittest," a phrase that philosopher Herbert Spencer coined and Darwin embraced, can easily misrepresent the idea, suggesting either self-assertion (the survival of the fiercest) or accommodation (the survival of those that fit in).

What's really meant is the survival of those selves that fit their circumstances well enough, in other words, the survival of the well fitted. Fittedness is tailoredness or suitability of selves for survival, *given, about*, or *with reference to* their circumstances.

One look at a tailored suit and you can tell something about the self it is tailored to fit. The suit is representative or *about* both the self that wears it and that self's context, perhaps a business context, but setting aside business culture, it is about the warm-blooded self's skin in the game, the aim to survive that is threatened without thermal insulation.

Were it not for selves and the circumstances in which they must survive we would expect a more haphazard arrangement of materials. The same goes for the skin that selves exhibit in the wild. For example, feathers too are tailored to fit the self to its circumstances. Signs of life on the planet or in a clothes closet provide evidence of traits that suit selves given, about, or with reference to their circumstances.

Inanimate artifacts like suits, tools, and machines are fitted to circumstances by the work that selves do to achieve their aims. A computer fits the user's aims, given the user's circumstances. A nest fits a bird's aims given the bird's circumstances. The river altered to irrigate crops is suited to fit farmers' aims, given their circumstances. As such, the fittedness fits selves to circumstances and reflects, represents, or is *about* both the self and its circumstances. From functional or fitted features, we can tell something about a self's aims and circumstances. I will call this quality *about-ness.* About-ness only occurs for selves.

In the absence of selves and aims, nothing is about anything. A crack in a rock might tell us something about the rock's history, but the crack doesn't make the rock fit its circumstances in ways that are significant or valuable for the rock given its aims.

About-ness is a distinguishing feature of selves. Drives, desires, intentions, and all the rest of the qualities that we associate with aims are always *about* circumstances for selves in a way that cause-and-effect events are not. As biologist Jesper Hoffmeyer argues, "Thoughts, hopes, desires etc. are always about something else, and we distinguish them according to what they are about. Lifeless things, on the other hand, such as stones or clouds are not—to the best of our knowledge—about anything else."[1]

GENERATING VS. TAILORING TRAITS

We humans tailor our artifacts to our aims and circumstances, but who or what does the tailoring for a self's biological adaptations? Not natural selection. It is no designer prototyping traits and selecting the best for rollout through survival and reproduction.

So instead, do selves do the tailoring? To answer accurately, we have to distinguish between generating functional traits and tailoring them. Tailoring implies a proactive effort to evolve improved adaptations.

We humans are learners, proactive tailors seeking ways we can fit our circumstances with greater efficiency and effectiveness. We aim for better fittedness. We even can have an aim to learn better ways to aim harder to learn, to stay motivated to learn faster, more efficiently and productively.

But the vast majority of selves don't learn. They are adapted but they don't actively adapt. Unlike humans, most selves don't tailor their own traits. A plant doesn't aim to evolve more productive foliage. A bear doesn't aim to evolve fur for more efficient thermal insulation.

Still, all selves generate their own functional traits, traits that are valuable and significant *for* themselves *about* their circumstances. Even the first selves to emerge from chemistry anywhere in the universe would have the capacity to generate their own functional traits.

Neither chemistry nor natural selection tailors traits, and selves without the capacity for learning don't tailor their own traits either.

People often intuit that traits are a product of natural selection's tailoring. Alternatively, people often intuit that selves aim to tailor their own traits the way humans do through learning. The vast majority of traits are adaptive but not tailored, meaning functionally fitted but not by anything aiming to make them fit.

Natural selection and survival of the fittest are best understood as *reduced reproductive success of selves poorly fitted to their environmental circumstances*. Neither individual circumstances nor the environments are selves. Circumstances may contain selves such as predators and prey, but the predators do not aim to tailor prey and prey do not aim to tailor predators.

In sum, all selves generate their own functional, fitted traits, which are *for* themselves and *about* their circumstances. Nonselves don't generate their own fitted traits. At the origin of selves and throughout most of evolutionary history selves have not aimed for improved fitness, and natural selection doesn't aim to improve fitness either. Only with a capacity for learning do we see anything like adaptation or active tailoring as a true aim to achieve improved fittedness.

TRAITS ARE ABOUT SOMETHING FOR SELVES

If all selves generate their own functional traits without, in most cases, tailoring them, how can we tell what is and isn't a functional trait? We might be tempted to define a trait as anything that we can imagine serving some functional use, but that would encourage us to animate nonliving phenomena. We can always make up a reason why something might be functional, the water flowing downstream, for example, treated as a trait functional for getting water to the ocean.

Instead, what makes a trait functional is a three-way relationship linking the trait to self to circumstances by means of for-ness and about-ness:

A functional trait is significant or valuable for a self about its circumstances.

Two clarifications: First, so far I have mostly treated selves and aims as distinct phenomena. They can be distinct but they aren't at the origin of selves. Selfhood emerges as the aim to survive and reproduce, in other words, the capacity for *self-regeneration*, the self's defining aim about which more shortly.

To flesh out the triadic relationship we could say that *a trait is of significance or value for a self (given its aims) about its circumstances*, but we can avoid this cumbersome explicitness if we simply assume that a self is also its aims.

Second, selves generate their own traits. Traits are therefore *by and for* selves about their circumstances. We can contrast this with what we find in machines and other artifacts. A pacemaker is designed by selves to generate electrical pulses of value or significance not for the pacemaker but for the selves that use it.

CAUSED EVENTS VS. INTERPRETATIVE BEHAVIOR

Recognizing the triadic relationship of traits, selves, and circumstances will prove useful as we next explore the true nature of information, a three-part relationship coupled by for-ness and about-ness: *events are of significance about circumstances* for *selves, given their aims.*

About-ness is representation—the way that a trait suits and therefore represents a self's circumstances. According to Deacon, "the riddle of life is ultimately the riddle of representation and how it could have spontaneously emerged from nothing but chemistry."[2] Representation is not a chemical cause-and-effect relationship. Rather a representation results from *interpretation*, a kind of trait common to all selves.

The self's relationship to information is where for-ness and about-ness are most apparent. Information is always *significant for* a self *about* its circumstances. Interpretation is not a simple two-part cause-and-effect relationship but a triadic relationship whereby a *self* (1) interprets an association between *a sign* (2) and *what it's about* (3) for the self.

We often lose sight of the role played by interpretation. We do so by treating information as completely contained and produced within some material object. For example, we might say that a book contains information that causes us to think certain thoughts, that a stop sign contains information that causes us to stop, that pheromones contain information that causes animals to mate, or that DNA molecules contain information that causes adaptive traits.

Treated this way, information invites equivocation whereby on the one hand we can say that the sign is a material cause of material effects—the information within an object causing a self's action, and on the other hand we can say that the material object aims to convey information to us.

Treating information as cause of effects lends false credence to eliminativism. Information is often treated as a property of a material object that causes us to act. Treating it this way promotes the false notion that all phenomena are reducible to cause and effect.

Or we can equivocate in the opposite direction, treating information as homunculi, "little men" inside an object aiming to tell us something, for example, books and stop signs aiming to inform us, or DNA aiming to program bodies. Treating objects as aiming to convey information lends false credence to *panpsychism*, the theory that all matter, living or nonliving, has aims in mind. A panpsychist might argue that if signs are aiming to convey information, then there are aims in anything and everything.

We also often say things like "I interpreted this as a sign that I should act." This is a more accurate way to think about information. A stop sign doesn't cause us to stop unless we crash into it. A stop sign is instead a *potential sign* that some selves interpret as signifying that they should stop.

A stop sign is not inherently information. Plants, animals, and people from distant cultures don't interpret stop signs as about anything. Potential signs are open to various interpretations. For example, a stop sign, *for* you, might be interpreted as *about* traffic safety, not getting a ticket, too much government intervention, or slowing down to enjoy life. Interpretation is unpredictable in ways that cause-and-effect events are not.

Cause-and-effect events only occur through the interaction of present material objects—Thing X and Thing Y interacting. In contrast, even the absence of something can be interpreted as a sign. A missing stop sign might be interpreted as about vandalism or lax traffic safety. The absence of an expected RSVP from a friend can be interpreted, among other possibilities, as a sign that there's been a misunderstanding, that your friend didn't receive your invitation, or that your friend is uncaring.

The absence of a cause never yields an effect, but the absence of a sign can yield an interpretation, and not just for humans. For example, a deer tick waits in high branches for the scent of butyric acid wafting up off the coat of a passing deer. When it senses the butyric acid, it falls from the branches, with luck onto the deer's coat. The deer tick is patient, waiting in the same spot for about six months. If six months pass without any sign of butyric acid, the tick moves to a different location. The absence of butyric acid is significant to a deer tick, given its interpretative competence.

I'll be making a strong distinction between cause-and-effect phenomena and interpretation. Only selves interpret. We interpret potential signs. Anything in the entire universe is a potential sign, but only becomes an *interpreted sign* when selves interpret it as significant.

Since interpretation is a fundamental trait of all selves, I'll be exploring how it works in more depth toward the end of this book. Before that, in chapter 12, I'll be exploring information theory, but otherwise I'll steer clear of the ambiguous term *information*, focusing instead on *potential* and *interpreted signs*. Anything can be a potential sign. It only becomes an interpreted sign when interpreted by a self as a sign *for* the self, *about* its circumstances.

8

AIMS

AIMING AS CONSTRAINING

From the simplest single-celled bacterium to the most ambitious world leader, from life's earliest as-yet-undiscovered missing link from nonselves to selves to the hardest-hustling twenty-first-century entrepreneur, all organisms work to achieve their aims. Nonselves do nothing of the sort. Though formal definitions of life rarely list aims as a distinguishing feature, they are one.

Complexity doesn't distinguish life. A snowstorm rivals a bacterium's complexity, but a snowstorm has no aims. Size doesn't distinguish life. Galaxies are huge but have no aims.

We associate aims with work, the striving or effort that selves do to achieve their aims. One might assume therefore that what distinguishes aims is the work involved. This would be a mistake. Aims are not work, but rather how work gets focused, directed, or channeled—in a word, *constrained*.

Later we'll explore in depth the relationship between energy, work, and aims. For now, we'll make this distinction: Energy is the potential to do work. Work occurs when things interact that happen aimlessly in the realm of nonselves, for example, molecules interacting and doing work on one another.

Aimed work is energetic interaction that serves a self. Of all the work that could occur, in selves the range of work is constrained, limited, or restricted such that it is of value or significance for the self with respect to or about its circumstances.

A bacterium can distinguish between high and low concentrations of glucose and then aim for the higher concentration. The glucose provides energy for the bacterium to do work. Like any energy source, glucose can be used for a wide variety of work. For example, if you ignite glucose it burns, producing heat that, assuming enough glucose, could warm a house.

Of all the possible work that the glucose could do, within the bacterium those possibilities get channeled, limited, narrowed, constrained, or aimed into work that benefits the bacterium. Work isn't possible without energy, but glucose energy doesn't aim itself. The bacterium does the aiming. All selves have this capacity to channel or aim energy into work to achieve aims on their own behalf. We know a self's aims by the functionally constrained work it does. The aim is how the wider range of possible work is narrowed to work that benefits the self.

PINPOINTING VS. AIMING

There are two ways to think about aiming, one almost exclusively human and the other universal to all selves. When we humans aim, we can pinpoint. We can say, "I'm going to be at this precise address at that time," and then arrive "on the dot" though not exactly a dot in the Euclidian sense—not an imaginary zero-dimensional point. The address and time we pinpoint are actually a range, more like the whole area of a bull's eye, though with a somewhat fuzzy outline. Have you arrived on the dot if you're two minutes late? Three minutes early? Probably close enough.

Still, the idea of pinpointing has some resonance for us humans. We can easily and mistakenly assume that all selves have pinpointed aims in mind, as though the bacterium pinpoints the glucose.

Our tendency to assume aiming is pinpointing is a symptom of our tendency to assume all phenomena can be explained in terms of pinpointed causes and effects. This in turn reflects our capacity for language, which the bacterium doesn't have.

With cause-and-effect phenomena, individual physical state B inevitably follows from individual physical state A. There's no aim involved, just physical laws and highly predictable outcomes.

With language we humans can declare pinpointed aims, not that we ever really mean pinpoints. Without language the bacterium can't say, "That specific glucose over there, I'm aiming for it!"

The bacterium aims, but only we humans can name the targets of our aims. We can name them with varying degrees of accuracy, even identifying pinpoints. But even when we name a pinpointed target, the name applies to a range of possibilities, not some Euclidian point.

Even without language, the bacterium aims. When glucose concentrations cross a threshold, the bacterium's behavior changes. Rather than drifting anywhere it narrows or constrains its range of motion to the range that serves the self.

Of all the behaviors the bacterium could engage in, it engages in a constrained range. The work it does is the result of other work it doesn't do, the work that is constrained away. Aimed work is channeled or constrained into a restricted range of work, not a pinpointed aim, but a narrowed range of possibilities.

IF NOT DETERMINISM, CONSTRAINT

If the universe were exclusively deterministic, there would be no possibilities, because every event would be determined by prior events. In a probabilistic universe, the question becomes how everything possible doesn't occur.

Everything possible doesn't occur because of constraints, defined as reductions of possible paths making some paths more probable than others. Constraints channel wide possibility into narrower probability.

We know this from personal experience. To focus our efforts, we reduce distractions. To do anything deliberately, we *de-liberate* ourselves, reducing our freedom to do alternative things.

We can model a bacterium's behavior logically with words, using a simple if-then statement: "If glucose, then move to it." This logic conforms to our deterministic sense of causes and effects, with glucose as the pinpointed cause of the movement's pinpointed effect. Such modeling simplifies in gratifying ways, but to understand how the bacterium aims, we have to update our scientific perspective and focus on how possibilities get constrained down to aims.

Aims distinguish means-to-ends behavior from cause-and-effect events. What then are the bacterium's means, and what are its ends? The bacterium's means include its capacity to aim or channel that glucose into the bacterium's functional work, which among other things includes finding more glucose.

And the bacterium's most fundamental end? As I've already suggested, it's circular. The bacterium's fundamental end is regenerating its ability to aim work. It regenerates its own aims, and reproduces them, regenerating its aims in its progeny selves.

This circularity is common to all selves. We all channel work into regenerating our ability to channel work. Our fundamental means are our capacity to channel work. Our fundamental end is regenerating our capacity to channel work. How selves do this when nonselves don't is what we must explain to solve the mystery of purpose.

DETERMINISM VS. PROBABILITY

People tend toward simplistic, deterministic assumptions about how change happens. Determinism makes explanation and prediction easy, eliminating debate and doubt.

Whenever we experience a reliable tendency, we tend to collapse it down to a simple cause-and-effect account of what happens. Consumer optimism causes stock prices to increase, a neurotic parent caused the child's sour temperament, and water shortages caused the war—a pinpointed cause, a pinpointed effect.

Though we worry about the implications of a deterministic universe, the way it traps us within immutable fates, we also embrace determinism. It allows us to treat the world as reliable clockwork, more reliable if we had the capacity to determine every pinpointed cause of every pinpointed effect.

French mathematician-astronomer Pierre-Simon de Laplace epitomized scientific determinism with a conjecture based on Isaac Newton's discoveries of the reliable laws of motion. Laplace imagined an intellect with infinite calculating power (or, as it came to be known, *Laplace's demon*) that could pinpoint the position and direction of every particle in the universe and apply Newton's laws to them. Laplace declared, "For such an intellect nothing would be uncertain and the future just like the past would be present before its eyes."[1]

Deductive logic and mathematics were early sources of confidence in determinism. Both are "plug-and-chug" systems, more reliable than machines since the rules and axioms of logic and mathematics don't break down. One simply has to input variables and apply the rules to achieve reliable outputs. As we'll explore in chapter 28, there's nothing simple about inputting variables.

The trend in science over recent decades has been away from the sense that the universe is deterministic at its foundations or anywhere else. Newton's laws are reliable. We can predict what will occur with a high probability of proving right. But that doesn't mean that they're deterministic, especially when applied to dynamics, large populations of elements interacting.

In science by the mid-1800s, determinism was losing ground to probabilistic thinking. James Clerk Maxwell, who among many achievements formulated the classic theory of electromagnetism, marked the transition: "The true Logic for this world is the Calculus of Probabilities, which takes account of the magnitude of the probability (which is, or which ought to be in a reasonable man's mind). This branch of Math, which is generally thought to favour gambling, dicing, and wagering, and therefore highly immoral, is the only 'Mathematics for Practical Men,' as we ought to be."[2]

POSSIBILITIES VS. PROBABILITIES

With probability, many things are possible, but some are more probable or likely than others. The calculus of probability is a system for estimating relative likelihood, so the textbooks say, but we haven't attended fully to its implications for the emergence of selves and aims.

Materially and energetically, there's nothing new under the sun. Conservation of matter and energy are reliable truths about the universe. Down to the scale of subatomic particles, the total of all matter and energy is present and conserved forever. New particles are neither introduced into the universe nor removed from it.

Still, one look around you, and you'll spot many artifacts that are new, artifacts that were highly unlikely billions, millions, or even thousands of years ago. And it's not just artifacts but the selves who create them to serve their aims. If any of these items suddenly appeared under the sun before life, it would be miraculous, but not because they're made of new matter introduced into the universe.

If we and our artifacts are not the products of new matter, what accounts for these undeniably new things under the sun? Was it a change in the laws of nature?

No, there are no new physical laws under the sun. Set aside what happened in the first three seconds of the universe or what preceded the big bang if anything. The laws of physics and chemistry have been consistent ever since. The theory of everything pursued by physicists is the theory of *everywhen* as well.

If physicists discover additional laws, they will only be new to us. They will not be laws that began operating only at a particular point in time. Physical laws hold good throughout all time.

If neither new matter nor new physical laws were introduced into the universe, how can we explain all of the new objects under the sun? We can only do so through a change in relative probabilities, some dynamic possibilities becoming more likely as a result of other possibilities becoming less likely.

How, then, do probabilities change, and how in particular did probability change such that selves and aims became extremely likely on Earth, given that long before life they were highly improbable? The answer requires another look at how evolution occurs.

9

EVOLUTION'S LIMITED LIMITING ROLE

EVOLUTION IS A PROBABILISTIC SCIENCE

Since we gravitate toward evolution as the explanation for life, before we begin to solve the mystery of purpose, we should clear up some points about evolutionary theory.

It's possible to treat evolution as simple cause and effect, arguing, for example, that a need caused a trait, or that natural selection programs our DNA, hardwiring us like robots to behave in predetermined ways. But with a little reflection, we recognize that evolution is inherently probabilistic. Natural selection constrains dynamic possibilities.

It does so imperfectly. An adaptive trait does not guarantee success. Not smoking doesn't guarantee that you won't get lung cancer; it just reduces the probability. Stopping at red lights doesn't guarantee survival. We know that after we stop, it's possible a rear-ender will kill us. Life is guesswork. We all know that, even though we don't always talk as though we do.

Although it's often said that natural selection designs this or that trait to target a pinpointed function, selective pressure is no engineer zeroing in on a pinpoint. We can speculate about which aims a trait might serve, but we can't pinpoint explanations and natural selection can't either. Many nonexistent species with different traits could probably survive in any given environment.

Flat-earthers once thought that it would be possible to sail off the ambiguous edge of the earth. Ridiculous as this now seems to us, it provides a useful metaphor for how evolution works. Selves can fall off the edge of survivability. Selves accumulate the constrained traits that enable them to reduce the likelihood of falling off the edge. Within the boundary outlined by that faint edge, selves are free to roam, relatively unconstrained.

The edge of survivability does not outline a flat area in which life is equally viable, nor does it rise conically to pinpointable pinnacles of fitness. Natural selection defines the edge of reproductive success and therefore lineage survivability. Lineages do terminate, falling over that edge, but more often, natural selection shapes relative fitness of lineages, some proving more fecund than others. Evolution enables selves to explore for viability.

NATURAL SELECTION IS AIMLESS

Some assume that evolution programs us, but this is problematic. Biologists often remind us that evolution by natural selection has no aims. Natural selection is sometimes used to describe the overall evolutionary process. More accurately, it's the process of elimination that results from populations expanding exponentially when resources are limited. As the latter, it is one of Darwin's three principles, the other two being heritability and variation.

Selection is a notoriously distracting term for the process of elimination since only selves select to suit their aims and natural selection is not a self. As philosopher/psychologist James Mark Baldwin argued, "Natural selection is too often treated as a positive agency. It is not a positive agency; it is entirely negative. It is simply a statement of what occurs when an organism does not have the qualifications necessary to enable it to survive in given conditions of life. So we may say that the means of survival is always an additional question to the negative statement of the operation of natural selection."[1]

Darwin chose the term *natural selection* to draw a parallel to artificial selection, breeders selecting plant varieties and individual animals with traits that suited the breeders' aims. The analogy is helpful only if we remember where it breaks down. Breeders select; natural selection only results in differential reproductive success, some lineages proliferating more than others, and natural selection operates passively, with no aims in mind, only on proliferating populations of selves with aims.

To say that natural selection explains life is like saying that erosion explains mountains. Erosion doesn't explain how mountains rise, just how they're degraded over time.

ONLY SELVES AND AIMS EVOLVE

The beginning of time is the beginning of differential durability, some things lasting longer than others. Differential durability can seem like evolution. If one galaxy lasts longer than another, one might try to claim that this is the equivalent of Darwinian survival of the fittest under natural selection.

It is not. Rocks form and dissolve at different rates under different conditions, but that doesn't mean rocks evolve. Molecules form and split at different rates, but that doesn't mean that chemicals evolve. This creates a serious problem for the popular interpretation of evolution as reducible to the differential durability of DNA molecules, and, indeed, for any model that suggests that life starts when molecules begin copying.

Only with selves does relative endurance become differential survival and reproductive success in Darwin's struggle for existence. Only selves struggle to survive and reproduce prolifically.

And the priority is survival, not reproduction. A self must endure the same natural law tendencies that erode the rocks and molecules. A self cannot reproduce if it doesn't exist, and it cannot endure through simple durability. It must engage in ongoing self-repair: the healing, replenishing, and replacing that occurs in a body throughout a self's existence.

REPLICATION VS. REGENERATION

Evolutionary theory explains beautifully how selves and aims are honed to fit environments, but it does not explain how selves and aims emerged, or what they are. We must be clear about this because people tend to evade the issue: natural selection doesn't explain the genesis of selves and aims, just how they adapt over time and only in conditions where two other features are present. Darwin described these as heritability and variation.

Today many replicator theories of evolution gloss over the mystery of purpose by treating heritability as mere copying, and variation as imperfection within the copying. Simplifying like this can create the false impression that all there is to the origin of life is differential copying, for example, molecules replicating in a chemical chain reaction.

Replication is the mere copying of inanimate stuff. For example, *catalysts* can facilitate chemical reactions, thereby transforming *reactants* into molecular *products*, producing lots of replicas of the same molecule types. That's replication and it will continue for as long as there are available reactants, after which the chemical reaction will peter out.

One can set up a "competition" between catalysts, each "vying" to transform the reactants before the other catalysts does. One of the catalysts might "win" the competition. This may appear to be like evolution, but only to us as observers, imagining the catalysts as aiming to win. We could do the same with two balls rolling down a ramp, each competitively "aiming" to get to a finish line first.

In imagined chemical competitions there's really just a temporary proliferation of aimless molecules. It is temporary, because as the reactants are eventually depleted, the chemical reaction degenerates and the products dissipate, all as a consequence of the aimless second law of thermodynamics—the universal tendency for organization to become disorganized, including the tendency for chemical reactions to peter out.

Purpose is not the product of the differential survival of replicating chemicals, not even the replication of naked DNA or RNA molecules, the molecules that now play crucial roles in all known forms of life. Replication peters out, but selves haven't for billions of years. Selves somehow work against the second law, converting energy into aimed *self-regeneration*.

For selves and aims to have survived continuously these billions of years, given the second law tendency toward degeneration, selves must be capable of self-regeneration, the most fundamental and universal aim—a capacity to regenerate ourselves faster than we would otherwise degenerate given the second law.

While we can picture the differential survival of replicating chemicals as like evolution, biological evolution is different, and must have begun with the emergence of selves and aims, in other words, chemistry capable of self-regeneration. To solve the mystery of purpose will require explaining how self-regeneration emerges and works. For now, we need only note that Darwin's concepts—heritability and variation—only apply to heritable and variable capacities to self-regenerate, not to mere molecular copying or replication.

EVOLUTION VS. ORIGINS

After publication of *On the Origin of Species*, Darwin was challenged on his failure to address life's origin, most notably by his erstwhile detractor turned occasional and ambivalent ally, biologist Richard Owen, who took Darwin to task for closing his book with the statement that "there is grandeur in this view of life . . . breathed into a few forms or into one."[2]

In a review of the book, Owen characterized the implications of "breathed" as "Pentateuchal," meaning resonant with the first five books of the Old Testament, arguing, "the doctrines of the generatio spontanea [spontaneous generation—origin of life from nonlife] and of the transmutation of species [evolution] are intimately connected. Who believes in the one, ought to take the other for granted, both being founded on the faith in the immutability of the laws of nature."[3]

Darwin, a cautious researcher, retorted, "Is there a fact, or a shadow of a fact supporting the belief that these [chemical] elements, without the presence of any organic compounds, and acted on only by known forces, could produce a living creature?"[4]

More than 150 years later, it's understandable if we feel torn about this exchange. Evolutionary theory explains so much that, before Darwin's insight, could only be explained by God's guiding hand. With Darwin, so much of God's hand was lifted out of the picture that many scientists continue to think, with Owen, that God's lingering finger sparking life's origins should be lifted as well.

But cautious researchers today would still side with Darwin. Though we have evidence that organic chemicals can form, we still do not have evidence that they could spontaneously organize into living selves anywhere near as intricate as the simplest known organisms.

We who do take evolution for granted are inclined to take spontaneous generation for granted too, but, at present, we are forced to do so on faith, the assumption that someday we will be able to explain how selves emerge spontaneously in an abiotic environment. And on that faith, many assume the mystery is as good as solved.

Supporters of intelligent design claim that evolutionary theory can't explain the evolution of complex functional traits such as eyes. Evolutionary biology shows that in fact, it can, by means of myriad, incremental, accumulating refinements—Darwin's "numerous, successive, slight modifications."[5]

By focusing on seemingly improbable evolutionary transitions, supporters of intelligent design miss a far more vulnerable scientific target. Evolutionary theory doesn't explain how selves and aims emerge in the first place. Until we have a sound solution to the mystery of purpose, natural science explanations for our existence will remain vulnerable to this challenge.

III

DEAD ENDS, LIVE CLUES

10

THE HISTORY

AN OVERSTEP IN THE RIGHT DIRECTION

We might expect that we would be further along in our search for a solution to the mystery of purpose, but we've had a bit of a setback, an overstep in the right direction, science throwing the aiming baby out with the supernaturalist bathwater. Here, too briefly to do it justice, we'll survey the history of that overstep.[1]

Our story starts with Aristotle, who distinguished four tools of his explanatory toolkit, his *four causes*.[2] Aristotle didn't use a house-building metaphor to illustrate the four causes, but many do, and we will here:

1. Material cause: The materials from which a house is built;
2. Efficient cause: The nail-banging and board-sawing that result directly in a house. Think "effects" as in cause and effect or "efficacy";
3. Formal cause: The form to which a house conforms, the architect's drawings;
4. Final cause (in Greek, *telos*): "That for the sake of which something is done," for example, the house built as means toward the end of providing shelter for someone.

Aristotle assumed that all change required explanation from all four causes, final cause included. An acorn's end goal was to become an oak (he called that drive the acorn's *entelechy*, or "internal end"). Rocks fell to earth because they too had a telos, a final cause or end-directed urge to be where they belonged, as close as possible to the center of the earth and therefore the universe for geocentrics like Aristotle. To Aristotle, "Nature does nothing in vain."[3]

Of these four, material cause and efficient cause are the causes we witness as straightforward. We can touch material objects. We don't typically debate their existence because they're reliably and objectively apparent to our senses. Likewise, our senses provide us with daily evidence of efficient cause, the cause-and-effect interactive work between material objects. Appreciation of these two causes comes easy to us all.

When we think about formal or final cause we tend to do so by means of an analogy from material and efficient cause. In the house-building analogy, the form is in the design, a physical model of the house's shape. We can therefore think of formal cause as like a blueprint causing a house, one material object causing another to conform to it.

Formal cause is thus easiest to picture as some kind of form-endowed thing causing changes in other things, perhaps something like the way, in injection molding, a mold imposes form onto whatever material is injected into it. Likewise, people tend to think of final cause as a nonmaterial thing that causes an outcome, such as the house builder's aim propelling the house building or a pool player's desire to win causing the eight ball to move toward the corner pocket.

Aristotle assumed that all effects had prior *final causes*, yet he recognized that this presented a problem in that there must have been a first cause. He solved it by positing an unmoved mover, the cause for which there was no prior cause. Aristotle describes the unmoved mover in terms that suggest final cause, a Godlike being contemplating only the perfect, in other words, aiming toward perfection. Aristotle also declares that the unmoved mover is indivisible, a property we associate with material objects. Thus, we see in his interpretation of the unmoved mover our tendency to equivocate, positing an immaterial/material object that aims for perfection.

AGREEMENT, THEN DISAGREEMENT

Aristotle's works were not religious, but rather the seeds of *natural philosophy*, as science was first known. Natural philosophy attempted a rigorous, logical examination of physical reality while still assuming that everything required explanation by final cause or aims. It suggested that the first aims were to be found in the unmoved mover, a nonmaterial indivisible object that was also a supernatural self with the means to achieve aims.

For close to two thousand years after Aristotle, Western civilization continued to honor the assumption that all events had final causes, ends, or uses. Some had internal ends (entelechies) and some had ends that were external to them. For example, a hammer is useful not for its own ends, but for the carpenter's.

It was easy to believe that all events had final causes. Indeed, it was useful, given the Christian West's growing commitment to an all-powerful, all-knowing God who had aims in mind.

When the Library of Alexandria burned, Europe lost its few copies of most of Aristotle's works, but the Islamic world still had them. Faced with the material challenges of empire building, the Muslims put Aristotle's works to use, contributing to many technological and scientific innovations.

Europe rediscovered Aristotle's works in the interaction between Europeans and Muslims in Spain and during the crusades against Islam. Through an approach later called *Aristotelian Scholasticism*, Europeans attempted to make Aristotle's theories thoroughly compatible with Christianity, calling all aims ultimately God's. By this account, everything happened for a reason, not just for a material, efficient, cause-and-effect reason, but for a good reason, God's reason. All events were God's will, his means to his ends.

For a brief time, Aristotle's approach to causality was compatible with both Christian and Islamic thought. The Christians were confident that understanding the physical world was a fine way to understand God's ends. The Muslims were less concerned with how Aristotle's works might reflect Allah's ends than with how Aristotle's natural philosophy could address the practical challenges of maintaining and expanding the Islamic empire.

The compatibility between Aristotelian natural philosophy and supernatural faith didn't last in the Islamic world. Influential twelfth-century Sunni philosopher Al-Ghazali and others challenged the coherence of explaining behavior as a function of both material/efficient cause and Allah's aims.[4] To do so was double counting and ultimately disrespectful to Allah. Islam tilted and then fell hard toward *fideism*, faith in Allah's aims as the source of all behavior, driven by a sense that it is hubris for humans to think one can understand Allah's aims.

TELOS, A LOST CAUSE

European natural philosophy took up where Islam left off. The crusaders brought many writings from the ancient world back to Europe, including Lucretius's poem *De Rerum Natura* (*On the Nature of Things*), based on a theory developed by Democritus and Epicurus that everything was made of indivisible aimless particles. Arguing against religious superstitions about heaven and hell, Lucretius declared that we need not concern ourselves with supernatural aims.[5]

To Lucretius, the only real causes were material and efficient cause, the atomic equivalent of nail-banging connection and board-sawing disconnection, push and pull among atoms. He argued that if the gods exist, they don't influence our lives, which are ours to enjoy, attracted to the pleasures that appeal to us according to the material and efficient cause configurations of atoms that we are.

Lucretius's long poem was repellant to the mass of atoms known as the Catholic Church, but its core argument nonetheless crept and eventually swept through European culture, appealing at a time when natural philosophers were beginning to make extraordinary strides in cataloguing material and efficient causes, confirming Lucretius's intuitions.

European investigations of nature did not always confirm the religious account of God and his aims. Copernicus discovered that the earth wasn't the center of the universe. Newton's account of motion revealed that "heavenly bodies," once thought to be moved by distinctly supernatural laws, moved by the same natural laws that governed natural earthly motion.

A commitment to explaining all behavior solely in terms of material and efficient causes yielded an unprecedented bounty of scientific insights and technical innovations. As Islam resolved its double-counting problem by counting only what mattered to Allah, Europe resolved it by counting only material and efficient cause. Francis Bacon, who heralded the scientific age in the 1600s, argued that pursuit of final-cause explanations hampered science: "For the handling of final causes, mixed with the rest in physical inquiries, hath intercepted the severe and diligent inquiry of all real and physical causes, and given men the occasion to stay upon these satisfactory and specious causes, to the great arrest and prejudice of further discovery."[6]

Eliminativist approaches became an increasingly resonant scientific interpretation of reality. According to this view, Aristotle was wrong about formal and final cause. All events can be explained as due to material and efficient cause. Purpose, or final cause, was entirely unnecessary to explaining what happens, since, in dissecting any seemingly consequential behavior, one always finds only material things interacting according to the efficient-cause mechanistic laws science was rapidly discovering. Final cause was becoming a lost cause. Natural philosopher Baruch Spinoza argued that "All final causes are nothing but human fictions."[7]

GUYS OF THE GAPS

Henry Drummond coined the phrase *the God of the Gaps* for the human tendency to posit supernatural agency to fill in gaps in our understanding of how things work.[8] Scientists refer to gap-filling, selflike entities as homunculi. We posit these *guys of the gaps* above us (gods), inside us (souls or spirits), all around us (the spirit in all matter), or throughout us (collective consciousness).

With the post-Enlightenment tilt toward eliminativism, scientists have been like bounty hunters rewarded for exposing and dissecting homunculi to reveal that they're always an illusion. Lightning is not God's wrath but electrical discharge. Disease is not a spell cast by supernatural demons but the consequence of biochemical interaction. The mind is not a special force; it's neurochemical mechanisms.

Sure, there remain some explanatory gaps to fill, but scientists know better than to fill them with purpose-driven homunculi, and anyway, they won't be gaps much longer. A little more dissection and we will have expunged all homunculi. No more gods or guys of the gaps, whether we imagine them above, below, inside, between, or all around us.

11

EVOLUTIONARY THEORY'S ELUSIVE SELF

THE GHOST THAT IS BUT ISN'T THERE

Evolutionary theory is often treated as the homunculi hunters' biggest bagging, the surest evidence of eliminativism's inevitable success. Still, trying to locate the self and aims in evolutionary theory can be like reaching around to grab an elusive mole scampering through tunnels, appearing and disappearing. We find evidence that the mole is in one of three holes: natural selection, bodies, or DNA. But when we reach for it, we're told that it isn't there.

Most intuitively, evolutionary theory can be read as an argument that functional traits yield benefits that satisfy natural selection's aims. Darwin introduced natural selection as analogous to artificial selection, farmers breeding for traits that suit their aims.[1] If natural selection is analogous to artificial selection, then natural selection selects traits that achieve its aims.

Evolution by chance modifications eventually became generalized as "the law of effect."[2] Not aiming for anything, organisms happen to reproduce with variation. Natural selection then selectively winnows the most adapted from the variety. The law of effect argues that variation is blind to aims. The variety we find in offspring need not be aiming to survive so long as the *effects* of variation can be selected upon.

Systems scientist and psychologist Donald Campbell summarized the law of effect and evolution more generally as "blind variation and selective retention," arguing that it is "fundamental to all inductive achievements, to all genuine increases in knowledge, to all increases in fit of system to environment."[3] If bodies are blindly varying, and yet selectively retained, it follows that natural selection is what aims to produce survivable bodies.

But evolutionary theory also argues that natural selection has no aims. It is merely the universal tendency for selves to survive differentially, some selves within populations dying and others surviving and proliferating. Differential survival entails differential death, but natural selection is no active murderer. Rather it's simply the passive environmental conditions that selves must fit through their means-to-ends behavior if they are to survive.

If organisms are aimless variants and natural selection is an aimless killer, maybe there are no selves aiming for anything. This would be a strong case for eliminative approaches, troubling though it would be to the universal intuition that purposes, selves, and aims are real.

SELF-FISHING

Darwin admitted that he didn't know what enabled traits to pass from parents to offspring. In 1900, with the rediscovery of Gregor Mendel's work, scientists gained insights into the laws of heredity, the inheritance of functional traits, and therefore the aims that functional traits produce.

In 1953, Watson, Crick, and Rosalind Franklin discovered that DNA was the chemical that makes genetic inheritance possible. Two decades after their discovery, George C. Williams and others argued for a DNA-centered approach to evolutionary theory that was soon made hugely popular by Richard Dawkins as *selfish gene theory*.[4] Selfishness, of course, suggests both selves and aims.

In his book *The Selfish Gene* Dawkins dismisses the intuition that our bodies are the self. Bodies are "replicator vehicles,"[5] the book argues, and, like machines, pure cause-and-effect dynamics. Machines for whom or what? For DNA pursuing its selfish aims of self-replication. By this popular account, DNA acquired its selfish aims when it, or a precursor chemical, began to replicate, gradually accumulating replicator-vehicle bodies that were selectively retained because they helped selfish molecules achieve their aims.

According to Dawkins, "We are survival machines—robot vehicles blindly programmed to preserve the selfish molecules known as genes."[6] Dawkins elsewhere declares, "The genes are master programmers, and they are programming for their lives."[7]

Thirty years after the original publication of *The Selfish Gene*, Dawkins conceded that the term *selfish* may have been misleading and that it might have been better to title his book *The Immortal Gene*.[8] Immortal is no less homuncular than selfish. His revised view, then, is that DNA, a chemical, is alive and lives forever.

Dawkins regards human consciousness as something special, speculating, "Perhaps consciousness arises when the brain's simulation of the world becomes so complex that it must include a model of itself." He argues that, with consciousness, "We, alone on earth, can rebel against the tyranny of the selfish replicators."[9]

This may make for a great morality tale, but tyrannical, immortal, and selfishly immoral chemicals aren't a scientific explanation, as critics have noted. Still, the simple vivid clarity of his interpretation has made it the tacit popular interpretation of evolutionary theory, tolerated and even accepted by many biologists.

EVERYWHERE BUT ANYWHERE

Dawkins also states, "Nature is not cruel, only pitilessly indifferent. This is one of the hardest lessons for humans to learn. We cannot admit that things might be neither good nor evil, neither cruel nor kind, but simply callous—indifferent to all suffering, lacking all purpose."[10]

Here Dawkins could mean that callously selfish self-preservation is the immortal and amoral molecule's chief aim, that nothing matters to natural selection, or that the eliminative scientists are right more generally in their nihilist assertion that nothing matters to anything anywhere.

Of course, DNA is a chemical, so thinking of it as selfish is a stretch. Dawkins treats selfishness as an innocuous metaphor, innocuous in that whether you think of DNA as having selfish aims or as a chemical that copies by cause and effect, the consequences remain the same.

By this account, it does no harm to attribute aims to chemicals, and indeed yields pedagogical benefit, since it resonates with our intuitions. Interpreting genes as selfish makes evolutionary theory easy to understand, and it doesn't compromise the real story, which is that nothing matters to replicators, replicator vehicles, or natural selection. The self and its aims are just an illusion reducible to cause-and-effect dynamics.

Dawkins describes natural selection as "a blind watchmaker" to counter Bishop Paley's famous argument that God must design organisms the way a watchmaker designs watches to suit his aims.[11] Still, even a blind watchmaker is a self that aims to make watches.

Lack of foresight is likely what Dawkins means, an allusion to the law of effect, whereby variation is blind. But that leaves us wondering where the self is hiding. Are bodies aiming to survive? Are selfish genes programming bodies in order to achieve their molecular aims, or are they the source of blind and therefore aimless variation? Is natural selection a blind watchmaker that aims to engineer successful replicator molecules with their replicator-vehicle ornamentation, or is natural selection a cause-and-effect process that eliminates the unfit? And the unfit what? Unfit aimless molecules of DNA or unfit aimless replicator-vehicle bodies?

If there are no selves with aims, evolution is worse than monkeys banging away on typewriters. With enough time, one of the monkeys might produce *Hamlet*, but no aims would exist by which to distinguish its value from any other random manuscript produced.

12

INFORMATION ABOUT NOTHING FOR ANYONE

BITS

Many people assume that life began when DNA became information. Still, it's not presently clear how that happened.[1] By themselves, DNA molecules are just molecules. They're not inherently information in the sense that we normally use the term.

Normally, we think of information as about something for selves given their aims. There's relevant and irrelevant, accurate and inaccurate, good and bad information, always with respect to something the information is about, what the information represents in our interpretation of it.

Of course, there is beneficial and nonbeneficial DNA, but that's not an attribute of the molecules themselves. When we think of DNA as information, we assume it has all the characteristics that information has in its conventional sense. Genes are information about circumstances for selves given the self's aims.

It has been possible to ignore how DNA becomes about anything for selves in large part due to the way current information theory defines information. Information theory originated in a technical report in 1948 by Bell Labs scientist Claude Shannon.[2] Shannon proposed the *bit* as a unit of measure for the quantity of communication that passes through a channel. The bit measures *reduced uncertainty*. The basic unit is a binary bit, the reduced uncertainty when one of two equally possible communications is received.

You ask a friend a yes-no question. Before he responds, you expect one of two possible answers, say, with a fifty-fifty likelihood of either. Once your friend has answered, the two possibilities are reduced to one actuality. That's a binary bit of information.

Bit measurement is easily applied to any kind of communication. For example, in the game Twenty Questions, the goal is to identify any possible object in the universe by a process of elimination, reducing uncertainty with no more than twenty binary bits of information. To play, you don't have to assume fifty-fifty probabilities. Still, you try to ask yes-no questions that each eliminate half the remaining possibilities, which is why we don't tend to start with a question like "Is it an article of clothing?"

Shannon's bits can be used to calculate nonbinary possibilities, such as a reduction from thirty possibilities to one actuality, or to calculate uneven odds, such as when a yes is far more likely than a no, or to calculate nonbinary reductions that don't yield a single outcome, such as when seventeen contestants are reduced to four finalists, each with different odds of winning.

Shannon's theory can be applied to how odds change with every step in a communication. For example, when, in his texted response to your yes-no question, your friend types a "Y," the chance that he or she will follow it with "es" becomes more likely. At every step of the communication sequence, likelihoods change in ways that can be measured in bits.

In developing his unit of communication, Shannon deliberately set aside the question of semantics or meaning, arguing that

> The fundamental problem of communication is that of reproducing at one point either exactly or approximately a message selected at another point. Frequently the messages have meaning; that is, they refer to or are correlated according to some system with certain physical or conceptual entities. These semantic aspects of communication are irrelevant to the engineering problem. The significant aspect is that the actual message is one selected from a set of possible messages.[3]

IF A BIT FELL IN THE WOODS

We can calculate the bits of communication even when no one is sending it. If there's a 50 percent chance of rain today and it rains, that's a binary bit even though no one sent the rain as a communicated message. The rain would still be information to someone who receives it.

But suppose it rains and there's no one there to receive it as information. Is it still information? It's still a binary bit, so according to information theory standards today it would still be information.

By this standard, DNA is inherently information too. Genes are strands of nucleotides. Nucleotides are of four types, A, G, C, and T; each is equally likely in any location on the strand. A one-in-four reduction of uncertainty is two bits. In other words, one could determine which nucleotide is occupying any location on the strand by asking two yes-no questions (for example, *Is it either A or G?* No. *Is it C?* No. *Then it's T.*)

Indeed, by this standard, any change can be measured in bits, so any change can be called information. In a lifeless desert, if a grain of sand falls in either of two directions and there's no one around to interpret it, it's still a binary bit of information.

Change happens all the time everywhere in the universe, and it can all be measured in bits. But few of us would really consider all of that change to be real information. It's all potential information, but it would only become information when a self interprets it as such.

Shannon was cautious about extending his theory to encompass all information, arguing,

> The word "information" has been given different meanings by various writers in the general field of information theory. It is likely that at least a number of these will prove sufficiently useful in certain applications to deserve further study and permanent recognition. It is hardly to be expected that a single concept of information would satisfactorily account for the numerous possible applications of this general field.[4]

Still, the extension of his theory has taken hold, and not just among information scientists. Many of us think of computers as information processors or even generators of information. We think of hard drives as storing information and we wonder when computers will store and generate enough information that they become selves. Some suggest that selves are just computers.

A computer is a massive bank of cause-and-effect switches, each designed to flip binary bits. The switches are not selves. They have no aims, either individually or en masse. A terabyte is roughly eight trillion binary bits. That's a lot of potential information that could be significant to someone, especially when you consider that a single bit would measure the yes-no answers to such questions as "Have they launched a nuclear first strike?," "Will we address climate change before mass extinction?," "Will I die this year?," or "Will you marry me?"

Still, how much we care about a terabyte of information depends upon its interpreted significance. If a hard drive contains your life's work, you'll care about that terabyte a lot. If it contains nothing but your lifetime collection of junk mail, you won't care at all. Either way, it's still a terabyte of information according to current information theory.

The problem with information theorists treating bits as information was recognized early in the theory's development. For example, scientist/mathematician Warren Weaver said: "We have to be clear about the rather strange way in which, in this theory, the word 'information' is used; for it has a special sense which, among other things, must not be confused at all with meaning. It is surprising but true that, from the present viewpoint, two messages, one heavily loaded with meaning and the other pure nonsense, can be equivalent as regards to information."[5]

SACRIFICING VIABILITY FOR INVIOLABILITY

The inclination to treat bits as inherently informational is consistent with a tendency common to many approaches to addressing the mystery of purpose, a tendency that could be called *sacrificing viability for inviolability.*

People have long sought to discover inviolable truths about reality, absolute formulas that would enable us to make inviolable predictions, and therefore to direct our efforts, confident that we will achieve our aims.

Newton's laws provided some of the first hope that science could uncover the inviolable laws for predicting everything, including the behavior of selves. Indeed, people are especially interested in the behavior of selves, given how much our well-being depends on how selves behave. Newton implied that it wouldn't be easy when he said, "I can calculate the motion of heavenly bodies, but not the madness of people."[6]

When we find an inviolable law that covers some aspect of what selves do, we tend to latch on to it as though it explains all of what selves do. Shannon's binary bit is a feature of information, but it is not the whole story. It measures an important feature of potential signs, but it ignores senders, receivers, and most importantly the self's ability to interpret signs as about something, given a self's aims.

To call bits information is therefore to sacrifice a truly viable theory of viable selves to accommodate our appetite for inviolability. As philosopher John Collier says, "The great tragedy of formal information theory is that its very expressive power is gained through abstraction away from the very thing that it has been designed to describe."[7]

The bit is an extremely versatile unit of measurement. In fact, it's too versatile. This versatility has distracted many from uncovering how DNA becomes information in the conventional sense and more generally what a self is as distinct from a bank of switches, no matter how massive that bank is.

13

THE ENGINEERED GHOSTS IN OUR MACHINES

FUNCTIONALISM

In the debate over whether computers are intelligent, some argue that they must be, since they can outcompete humans on tasks that call for great intelligence. Others argue that because they only process signs, transforming them from one form to another, they are not intelligent independent of users.

Computers are vast quantities of switches toggling one another between two poles that we name zero and one. The more switches, the more we think they are becoming selves like us due to the increased functions they can handle for us. By this reckoning, if we sum a massive enough collection of switches, eventually a self emerges. I'll call this tempting but insufficient explanation *thresholdism*, the assumption that once cause-and-effect dynamics cross a certain threshold of quantity or interactions, a self simply emerges.

There's an air of urgency about the debate over computer intelligence, driven in part by fear that computer intelligence is going to outpace human intelligence and take on a life of its own. Taking on a life of its own suggests that intelligence is synonymous with selfhood.

To any of us born in the decades before the computer age, the industrial revolution may have seemed nearly complete. Many of us had no idea how much automation was still possible. Every day we discover new ways to offload our functional tasks onto computer technology. Present as we are for this second industrial revolution, the temptation to subscribe to thresholdism is greater now than ever.

The dawn of the age of machines displaced human power and horsepower. The computer age displaces human mind-power at an awe-inspiring if vocationally threatening rate. With a year's work, a few computer programmers can design software that replaces a whole international white-collar workforce forever.

We take comfort from the thought that there must be something that computers can't do in our stead, but it's hard to say what that would be, and anyway, some computer engineer somewhere has probably begun working on getting computers to do that too. If computers can replace selves and their functional efforts, why can't we say that computers are becoming selves?

Functionalism argues that we can. According to functionalism, if two things function equivalently, then there is no justification for calling them distinct.

SELFHOOD IS NOT WHAT TURING TESTS

The most famous case for functionalism is the Turing test, proposed by Alan Turing, a founding figure in digital computing. The Turing test was his response to the question computers naturally sparked: Could computers ever become human? Rather than answering the question, Turing suggested a way to answer it. If a human is unable to distinguish between conversation with a computer and conversation with a human, then the computer is functionally equivalent to a human.

An international Turing test is held annually. Science fiction, such as the movie *Ex Machina* (2015), commonly features humanoid computers that would pass a Turing test. Most stories with humanoid computer characters are cautionary tales about machines becoming as ambitious, intelligent, functional, and slippery in the service of their aims as human selves are.

There are plenty of reasons to doubt that the Turing test is sufficient to determine whether a computer is as intelligent as a human. For an obvious one, how much conversation is sufficient to demonstrate unconditional functional equivalence?

More importantly, intelligence is one thing; selves with aims are quite another. What would be the Turing test equivalent for discerning whether a computer was a self with aims? Whatever it is, it wouldn't be a test of intelligence. A worm is not exceptionally intelligent, but it's a self with aims.

CLUSTERED CONSCIOUSNESS

If more binary bit switches aren't enough to cross the threshold to self, there are other potential variations on the thresholdism theme. One approach is to argue that it's not sheer numbers of switches that bridge the threshold from natural machine to natural ghost, but how the switches are interconnected. For example, Integrated Information Theory (IIT) argues that conscious experience is measurable as connectivity, not massive numbers of switches but massive integration of them, switches that switch one another in clusters. To quote IIT pioneers Giulio Tononi and Christof Koch: "A corollary of IIT that violates common intuitions is that even circuits as simple as a 'photodiode' made up of a sensor and a memory element can have a modicum of experience. It is nearly impossible to imagine what it would 'feel like' to be such a circuit, for which the only phenomenal distinction would be between 'this rather than not this.'"[1]

It would be nearly impossible, they seem to argue, because a photodiode's experience falls below some threshold that is reached by having many more switch connections.

Similar arguments go back as far as Lucretius, or at least as far as philosopher Thomas Hobbes, who, in the Enlightenment, made an eliminative argument that we are indistinguishable from machines: "For seeing life is but a motion of limbs, the beginning whereof is in some principal part within; why may we not say that all automata (engines that move themselves by springs and wheels as doth a watch) have an artificial life? For what is the heart, but a spring; and the nerves but so many strings; and the joints but so many wheels giving motion to the whole body."[2]

Philosopher Charles Taylor goes so far as to say of the distinction between life and machines that "Few biologists today think it worthwhile to pay much attention to that distinction."[3] Biologist Richard Lewontin criticizes this trend, arguing that "the ur-metaphor of all of modern science, the machine model that we owe to Descartes, has ceased to be a metaphor and has become the unquestioned reality: Organisms are no longer like machines, they are machines."[4]

We can always imagine a self's means-to-ends behaviors as the effect of switches, for example, that selves are programmed with the algorithm "If hungry, then eat." The problem is that we can also imagine any cause-and-effect behavior this way. For example, water flowing could be said to act on the algorithm "If path A is lower than path B, take it; if path B is lower than path A, then take it instead." Photodiodes likewise—for example, a dawn-to-dusk light fixture—could be said to act on the functional algorithm "If the sun is bright, then stay off; if the sun is not bright, stay on." Water flow and light fixtures can thus be read as functioning like selves with aims, but are they functionally equivalent in all respects?

AMNESIC WATCHMAKER SYNDROME

For all we know, computers will eventually become selves with aims, but no computer today is even close. To assume they are close is to fall prey to what could be called *Amnesic Watchmaker Syndrome.*

Like watchmakers who produce watches to serve their aims, human engineers design and build computers to serve ours. Forgetting this, we sometimes treat computers as though they were selves independent of our aims. Engineers who would argue that the computers they design are true selves with aims would be like blind watchmakers who are blind to their own watchmaking.

There's amnesic watch repair too. Our functional tools only remain functional because we repair them. Selves repair or heal themselves; tools and machines do not. We design durability and failure-prevention mechanisms into our tools and technology and in some cases modest capacity for self-repair. But nothing designed selves to be fail-safe. And though we can say that fail-safes and repair capacities evolved in all known selves, we can't say they evolved into the very first selves, the selves that must have emerged for evolution to even begin.

Computers are made of durable materials and have few moving parts. This is a fundamental source of their reliability. Still, reliability is not the same as self-reliance. Rocks are reliable in their behavior but no one would call them self-reliant.

Self-reliance is never absolute. All selves are islands of autonomy, but only because they are capable of importing the energy and resources that they need in order to stay alive, repairing, protecting, and reproducing themselves. Computers depend on external sources of energy also, but we have to plug them into power sources that we also design and produce to suit our aims.

DYNAMIC SYSTEMS THEORY

ITT is but one example from an expanding range of research topics, all of which contribute to our sense that science may have already solved the mystery of purpose through a combination of thresholdism and amnesic watchmaker syndrome.

These research topics include systems theory, cybernetics, dynamical systems theory, chaos theory, artificial life, self-organization theory, cognitive science, cellular automata, and complexity theory, which I'll generalize here as *dynamic systems theory*. All of these fruitful research programs focus on detailed understanding of dynamic patterns that result from the behavior of large populations of elements interacting.

Dynamic systems theory yields important insights that, like Shannon's binary bits idea, will be essential stepping-stones to a scientific solution to the mystery of purpose. Still, and contrary to common expectations, the insights fall short of explaining or explaining away selves and aims.

To illustrate the complexity theorist's general approach, let's visit John Conway's Game of Life,[5] a pioneering project in cellular automata research. Conway programmed a computer to present a matrix of square lights, each square surrounded by eight other squares. The simulation steps through iterations, an attempt at a loose analogy to life going through generations. Lights switch on and off based on four simple cause-and-effect rules:

1. If a lit cell has fewer than two lit neighboring cells in one step, the lit cell becomes unlit in the next step, a loose analogy to dying from underpopulation—a lack of supportive neighbors.
2. Any lit cell with two or three lit neighbors stays lit in the next step, in loose analogy to having sufficient neighbors.
3. Any lit cell with more than three lit neighbors becomes unlit in the next step as though dying from overpopulation.
4. Any unlit square with exactly three lit neighbors lights up in the next step, as though revived by arriving in an ideal life-supporting environment.

With this simple setup, and a few carefully selected beginning configurations, the cell matrix steps through lighted patterns that appear lifelike, as gliders swimming across the screen, shooters spitting out bullets, or even offspring, complex patterns that given our tendency to read aims into cause-and-effect behavior are easy to interpret as representing selves with aims.

GAMING THE GAME OF LIFE

Dynamic systems theories get much more complex than Conway's Game of Life, but most play on a general thresholdist theme: *simplexity*, complex lifelike forms resulting from the interactions of large populations of simple nonliving elements in strictly cause-and-effect interaction over time, always with the amnesic watchmaker tendency to ignore the programmer's aims in configuring the simplicity that generates the complexity.

The Game of Life is as close to deterministic as engineers can make it. There's nothing unpredictable about the iterations in Conway's flashing lights. From the starting configurations, all future steps in the program are entirely predetermined. We shouldn't mistake the observer's surprise at the resulting patterns as evidence of life.

Nor should we mistake proximity for functional dependence. Conway programmed his game of life to reflect neighbors cooperating and competing with one another in their effort to stay "alive" (that is, lit). Neighboring cells are programmed to turn one another on and off in various combinations. Still such mechanical switching does not make neighboring lights means to each other's ends.

The complex patterns that result from simple beginnings in Conway's game are no more end-directed than the pixels of a computer-animated movie watched on an LED monitor. We're impressed by the resulting cartoon, but we know we're just watching pixels, juxtaposed just so by the animators with aims. Animated film producers and viewers don't suffer amnesic watchmaker syndrome. But we somehow tend to suffer it a bit more in our response to dynamic systems theory models.

THE REVERSE-ENGINEERING FALLACY

Amnesic watchmaker syndrome is a symptom of a false assumption that science and engineering can be done in the same order, a mistake that could be called the *reverse-engineering fallacy.* Ideally, in scientific investigation we first describe, then explain, and then prescribe. Garbage in, garbage out—if we get our descriptions wrong, then our explanations and prescriptions will likely be wrong too.

Engineering reverses the sequence, starting not with descriptions of what is but with prescriptions of what the engineer aims to create. Starting from a prescription, the engineer works to describe a specific mechanism by which the prescribed function can be achieved and does so by means of explanations for how things work—their engineering know-how. As long as their engineering know-how is accurate, the engineers are free to describe a range of equally functional ways to achieve what their prescribed product is meant to achieve.

Thus while science moves sequentially from description through explanation to prescription, engineering move sequentially from prescription through explanation to description.

For example, consumer demand prescribes smart phones. Engineers, well versed in explanations for mechanistic behavior (for example, the laws of physics and chemistry), translate the prescription into a variety of possible descriptions (designs) for smartphones.

The reverse-engineering fallacy is the assumption that science can be done in the same order as engineering—prescription toward description, for example, as though engineering a simple robotic mind to suit our aims describes how simple animal minds work.

We are often reminded that "the map is not the territory"—our description is not the same as what we aim to describe. Of course the map is not the territory, but this doesn't mean that all maps are equally inaccurate. In general, inaccurate maps result from a tendency in science to start from a prescribed function the way engineers do and build back to a map or description of what could achieve that function. The reverse-engineering fallacy and amnesic watchmaker syndrome combine to give us the impression that Conway's Game of Life, or any other designed computer simulation, reveals how selves and aims operate.

Engineers design computers to be as close to deterministic as possible, distinguishing between zeroes and ones without error. Computers are designed to be as isolated as possible from outside influences—fluctuations in temperature, input voltage, air pressure, movement, or any other external factor. It's ironic, then, that people often treat computers, the pinnacle of engineered determinism, as functionally equivalent to selves, the most indeterminate systems we know.

14

SMALL IS DUBIOUS

QUANTUM SELVES?

These days we're witness to a rush of insights in quantum mechanics that loosely parallels the rush of progress in information science. Both generate marvels that impress people as possibly relevant to solving the mystery of purpose.

Contrary to popular conception, quantum mechanics isn't anything-goes mysterious. As physicist Sean Carroll says, "We understand an enormous amount about the theory—otherwise we wouldn't be able to make those predictions that have been checked to amazing precision. Give a well-trained physicist a well-posed question about what quantum mechanics predicts in some specific situation, and they will come up with the uniquely correct answer."[1] Still, there's a fundamental difference between how probability works in classical and quantum mechanics. In classical mechanics, you may not know the state that something is in, but it is in a specific state. For example, you may not know whether a coin is heads or tails but it's one or the other.

In quantum mechanics, there's *superposition.* A thing isn't in a particular state until we observe them, though not necessarily because we have observed. Unobserved phenomena can be accurately described only with a wave function, a thing being in all possible states at once. The wave function represents the weighted distribution of the all of the states it's simultaneously in. It's as though an unobserved quantum coin is not either heads or tails, but fifty-fifty both at once.

Observing quantum phenomena is somehow mysteriously correlated with the wave function collapsing, the likelihood of it being in a particular state peaking close to 100 percent in one state, which is why we then observe it in that state. Before observing the quantum coin, it is half heads, half tails, occupying both states at once. After observing it, the wave function doesn't collapse to flatness, equal distribution across all possible states. Rather, the probability distribution peaks nearly completely at the one observed state.

Superposition is all too easily analogized to our experience with selves, for example, a self being of two minds until it acts. We might imagine that if Newton were alive today and updated on superposition, he would say that he could calculate the motion of heavenly bodies, but neither the madness of people nor quantum phenomena, implying perhaps that they are related.

Some quantum theorists have suggested that quantum effects are truly observer-dependent, the act of measurement altering what occurs at the quantum scale. We've all experienced observer effects in our interactions with other selves but not with other inanimate objects, so when quantum theorists speak of observer effects, it's easy to get the false impression that quantum mechanics researchers will discover that quantum particles are selves with aims of their own, altered by observation. This is not what quantum physics itself suggests, or what most quantum physicists would argue.

And then there's *quantum entanglement*, or what Einstein rejected as "spooky action at a distance,"[2] the way a change in one particle can instantly change a formerly partnered particle even miles away, an effect that, again, can lead to the false impression that there's something like telepathic communication, or information passed between selves, an interpretation that, again, no serious quantum physicist would suggest.

The transition from quantum to classical mechanics occurs with the collapse of the wave function, a transition from fundamental ambiguity to unambiguous outcomes, particles in one position or another, measurable in Shannon's binary bits. If one thinks of this as information condensing into material objects, one can arrive at the argument that quantum information theory pioneer John Wheeler described as "it from bit"—in other words, a material object ("it") results from, in effect, quantum yes-no decisions ("bit") about what state to occupy. Wheeler concludes, "What we call reality arises in the last analysis from the posing of yes-no questions."[3]

Which selves pose these questions is not specified. Wheeler's framing suggests a chicken-and-egg problem: If reality arises from questions and questions are only asked by selves, how can questions be asked before there are real selves to ask them? As Deacon describes the problem:

> If things aren't real until we observe them, then our own existence as things in the world is held hostage to our first being able to observe ourselves. My world may have come into existence with my birth and first mental awakening, but I feel pretty sure that no physicist imagines that the world only came into existence with the first observers. But what could it mean for the universe to be its own observer irrespective of us?[4]

A NEW FLOOD OF WHAT'S ALWAYS ALREADY EVERYWHERE

In his best-selling book on information theory, *The Information: A History, a Theory, a Flood*, James Gleick admiringly describes the quantum information theory conclusion: "The whole universe is thus seen as a computer—a cosmic information-processing machine."

If the whole universe is a computer, whom is it computing for? We observers ask Wheeler's yes-no questions, so is the universe computing for us? We are in the universe, so is the universe a computer computing for a part of it? Or is it computing for some higher self outside the universe for which the universe's calculations matter?

Gleick's book is a masterful account of the history and status of information theory that nonetheless reflects current ambiguity about whether information is merely bits or is actually informative for a self about its circumstances given its aims. The phrase *information about* only appears seven times in this 550-page book, and *information for* doesn't appear at all.

Toward the end, the book transitions from the quantum argument that everything is and has always been information to an exploration of our modern-day information flood, without noting, let alone resolving, the obvious quandary: If every quantum event in the universe has always been information, how could there be an information flood now? How could there suddenly be more of what was always already here?

We really do experience an information flood these days, but not because there is increased production of cause-and-effect events measurable in binary bits. There aren't more potential signs in the universe than there used to be but more selves with more evolved and learned capacities to interpret these potential signs. One can't explain information without explaining the selves that interpret it.

Thus, as computer scientist Joseph Goguen points out, "It is said that we live in an Age of Information, but it is an open scandal that there is no theory, nor even definition of information that is both broad and precise enough to make such an assertion meaningful."[5] In chapter 25, I will introduce Deacon's approach that makes better sense of information.

LOOKING FOR THE KEY WHERE IT'S DARKEST

For all the potential some people see in quantum mechanics to explain consciousness, we can harvest very little from current quantum mechanics relevant to selves and aims. There are no phenomena in quantum mechanics aiming to regenerate themselves.

Perhaps future research in quantum mechanics will reveal phenomena that will suggest clues to contribute to a solution to the mystery of purpose. After all, the field is relatively new. It has proven to be a fecund source of surprises, generating as many or more new mysteries as new solutions to old mysteries.

Of course, if the mystery can't be solved through classical physics, that is, at the scale at which life is lived, then the exploration may need to be sought in the more esoteric and mysterious realms of physics. If, in the future, it comes to that, by then we may have better explanations for quantum behavior that will bring rigor to the exploration. Deacon's proposed solution to the mystery of purpose suggests that it need not come to that.

Searching for the key to life's distinct characteristics within the quantum realm is searching where it's darkest and most conducive to impressionistic, equivocal speculation. The dark, of course, has its appeal. Like a hypothetical supernatural reality, it is highly open to interpretation while remaining highly resistant to disconfirmation.

Of course, sometimes we must search where it's darkest because that's where the key is to be found. As we will see with Deacon's account, the key appears to be available within classical physics and chemistry just where we would expect it to be given that practically everything about bodies conforms to the already-well-illuminated laws of physics and chemistry at the living, not the quantum, scale.

FLOODLIGHTING THE NONSOLUTIONS

Are selves ghosts or machines? Here are four basic answers that we have touched upon in our exploration:

1. Supernatural ghosts (supernaturalism): The self is a supernatural thing, a soul or spirit, perhaps breathed into us by a higher supernatural self.
2. Just self dynamics (eliminativism): Selves and aims are illusions, entirely reducible to cause-and-effect dynamics.
3. Just machines (functionalism): Selves are like computers, functional machines, their aims programmed in as information.
4. Homunculi (natural ghosts by equivocation): So long as we can predict the consequences of behavior by positing selves and aims, we need not wonder about the origins or nature of selves and aims.

With the mystery unsolved for millennia, researchers have generated many variations on these themes, including the following five. Each of these approaches falls short of a true solution to the mystery of purpose:

Panpsychism: Selves and aims are all-pervasive in the universe. Even the smallest subatomic particle has aims.

The teleological hypothesis: In his short book *Mind and Cosmos*, philosopher Thomas Nagel makes a compelling case, similar to Deacon's, that natural selection does not solve the mystery of purpose. Nagel suggests that telos or purpose "may be determined not merely by value-free chemistry and physics but also by something else, namely a cosmic predisposition to the formation of life, consciousness, and the value that is inseparable from them."[6] In other words, there might be some fundamental feature of nature that makes selves and aims more likely than it would seem from mere physics and chemistry. Nagel, who reaches this conjecture by a process of elimination of alternative explanations, acknowledges that it is not a scientific explanation.

Universal Darwinism: In this view, whenever there is variation with differential persistence there's evolution. Everything evolves, including universes, inanimate matter, and selves. By this account, since evolution explains selves, universal evolution solves the mystery.

Multiversalism: This approach suggests that our universe is one of a near-infinite number of universes, so there's no need to explain what's unusual about ours.[7] Why are there selves and aims in our universe? Why not? Everything is possible in one universe or another.

Mysterianism: It is beyond humans to solve the mystery of purpose. Mysterianism is sometimes championed explicitly, though, more often, it is implied by rejection of all proposed solutions. Philosopher Jerry Fodor tilted toward mysterianism when he wrote, "Why is there anything except physics? . . . Well, I admit that I don't know why. I don't even know how to think about why. I expect to figure out why there is anything except physics the day before I figure out why there is anything at all, another (and presumably related) metaphysical conundrum that I find perplexing."[8]

Solving the mystery of purpose is obviously daunting. It has remained mysterious for millennia and even with all the scientific breakthroughs of the past few centuries. Even with the more recent revolutions in genetic, information, systems, and quantum theory, we seem no closer to a solution. What, if anything, could make it any more solvable now? The answer is a seemingly innocuous and largely unheralded yet pivotal scientific breakthrough of recent centuries. A shift in attention from what does occur to the relationship between what does and what doesn't occur opens a window to the heart of the problem and its solution.

IV

GROUNDING A SOLUTION

15

PROCESSES OF EMERGENT ELIMINATION

NATURALISM BUT NOT MATERIALISM

As we saw in our brief historical survey of attempts to solve the mystery of purpose, there is a very strong tendency for people to assume that all change results from some variation on Aristotle's material and efficient cause—material cause and effect—one material thing interacting efficiently or directly with another such that change results. This assumption is the basis for what's called *materialism.*

In the behavioral sciences *materialism* means a tendency to consider material possessions and physical comfort as more important than other values—not the meaning applied here. In philosophy and the physical sciences materialism means the doctrine that nothing exists except matter, its movements, and its modifications. Materialism is the basis for eliminativism, the argument that material cause and effect explains everything, or at least is the only kind of explanation that we should count as valid.

Materialism is irresistibly intuitive. It has been firmly and formally embraced since the scientific revolution, but we hear it even in Aristotle's original positing of final causes, for example, in his treatment of the prime mover as the immaterial yet material-like indivisible first cause of all effects.

Here I'll be pursuing solutions grounded in naturalism, but not materialism. Materialism and naturalism are often treated as synonymous. In a subtle yet crucial way they are different.

Nature includes absences—the absence of material in the spaces between material objects, but more importantly the absence of once-possible dynamic paths that have become impossible, and once-likely dynamic paths that have become unlikely. These eliminated or reduced possibilities are natural though not material.

A dynamic path is a way that dynamics might change over time, a possible dynamic flow or current. A simple example would be water flowing down a streambed. It's dynamic in that the water isn't standing still. It's flowing down a path, that is, in a general direction. Dynamic paths are also easily evident in car or pedestrian traffic flow.

It's easy to see these as dynamic paths because the flows are currents of moving elements—water molecules, cars, or people. But not all dynamic paths are that apparent. For example, a cup of hot coffee cools to room temperature by means of dynamic paths also, even though the coffee and cup molecules aren't traveling in currents. Rather, the speed at which the molecules are moving changes, forming heat currents visible with an infrared camera. The concept of dynamic paths applies to any energetic flow, for example, a change in pressure, electrical current, or radiation.

Here I'll be paying attention to changes in the likelihood of dynamics taking some paths compared to others. For example, without a riverbed to constrain paths, water could flow down various alternative paths. The riverbed limits or constrains the water flow to fewer of many possible dynamic paths.

A reduced probability of dynamics taking certain paths—in other words, a change in relative probability—does not have mass, volume, charge, parts, or any of the other attributes we associate with material objects. The natural world thus includes once-likely, now unlikely possible paths not attended to from the materialist perspective. Changes in likely paths are natural, just not material.

There have been things in the universe that are no longer things, and not by any violation of natural law. There are new things existent under the sun, but there's also the absence of formerly existent things and yet their absence is not a result of any departure from the natural realm. The dead are not likely to ever exist again, though not because their existence would violate the laws of nature, or because the materials they were made of have disappeared, or because they have been transported to a supernatural realm. Their probability of existing has changed. That change is purely natural, and not material.

Materialism expects and demands material causes for all material effects. If something changes, something must have interacted with it. In contrast, naturalism doesn't assume that all change is materially caused.

PROCESSES OF ELIMINATION VS. PROCESSES OF PRODUCTION

Changes in the relative probability of dynamics falling down one or another path can be initiated in either of two ways: *positively*, making some possible paths relatively more likely, or *negatively*, making some possibilities relatively less likely.

For example, if you want to increase your likelihood of achieving your aims, you can either work positively, trying to increase the chances that you will succeed, or negatively, trying to reduce the chances that you won't.

Positive and negative here have nothing to do with which is better, only with how probabilities change. Positive means making some possibilities more likely; negative means making some possibilities less likely. The positive approach yields changed probabilities by *processes of production*. The negative approach yields changed probabilities by *processes of elimination*.

Both have their place. Sculpting clay is mostly a positive activity. Sculpting stone is mostly negative, removing the stone that doesn't fit. Die-casting is positive; parts grinding is negative. To the extent that the artisan has a final product in mind, all of these activities are "positive" in the sense that the artist's aim to produce a sculpture is a process of production.

To us, highly proactive human selves, the positive approach is intuitive. You have aims. You work positively to achieve them. Reducing your chances of failure is just an often-overlooked side effect of positive success.

In trying to solve the mystery of purpose, people have long extrapolated from the positive approach, long assuming that something or some higher force made selves and aims more likely through a positive process of production. Even Darwin's reference to life as "breathed into a few forms or into one" suggests a positive process of production, something added to make things come to life. The materialist approach to explaining how things happen emphasizes processes of production. To explain what changed something, look for the material thing that produced the change.

In contrast, our approach here will be to explore how selves and aims might have emerged by a negative process of elimination, constraints that make some dynamic paths less likely, thereby making other dynamic paths more likely. Instead of asking what process of production produced life's self-regenerative dynamics, we'll be asking what process of elimination constrained away myriad alternative nonregenerative dynamics such that self-regenerative dynamics remained.

To Deacon, a living whole (an aiming self) is less than the sum of all possible dynamics. A self's capacity for self-regeneration results from the elimination of alternative degenerative dynamics. Deacon's key contribution is in cornering us with the mystery of purposes and then focusing on processes of elimination to address it.

Instead of asking what produces signal Deacon asks what reduces noise. Instead of focusing on how something happens, he focuses on how all the other possible things don't happen. Deacon's approach is in keeping with an overlooked trend in science that has yielded some of its biggest insights.

NEGATIVE SCIENTIFIC BREAKTHROUGHS

Darwin's breakthrough was in looking at what doesn't happen to explain what does. His insight came to him after reading Thomas Malthus on economic processes of elimination. Malthus argued that human populations expand exponentially whereas arable land expands arithmetically. To Malthus this meant that human populations are inevitably winnowed. Darwin realized that the same logic applied to evolution, a winnowing process of elimination that would be biased toward eliminating the unfit, resulting in fitter organisms remaining.

Information theory pioneer Claude Shannon also bucked positive intuitions in conceiving of information and communication as the products of processes of elimination. We tend to think of communicated messages as produced. We therefore naturally overlook the messages not sent. Shannon flipped this. The message communicated is what's left after a process of elimination removes alternative possible messages. A *bit* of "information" only makes sense when we pay attention to what is in relation to what could have been communicated but wasn't.

Dynamic systems theory also started with encouragement to think negatively. W. Ross Ashby, the pioneer in cybernetics who coined the term *self-organization*, remarked that "the cyberneticist observes what might have happened but did not."[1]

Still, processes of elimination are counterintuitive. It's simply easier to think of processes of production—positive causes of positive effects. Why does something change? We assume there must be a positive cause that yields the changes that we notice. We don't tend to think of something happening because other things didn't as often as we tend to think of something happening because something else did.

As with many counterintuitive breakthroughs, with time there has been a tendency to fall back toward intuitively positive interpretations. Ashby's systems theory focus on processes of elimination has largely dropped away. His choice to call the process self-organization doesn't emphasize the negative basis for it. Self-organization gives the impression of selves engaged in a positive process of production toward organization.

We see a similar trend toward positivity in the current interpretation of Shannon's theory, a trend that takes us from Shannon's treatment of communication as resulting from processes of elimination to the assumption that everything is positive information. If, as physicist John Wheeler suggested, "what we call reality arises in the last analysis from the posing of yes-no questions," then the cosmos is just an extended game of twenty questions with a universe that has positive things in mind.

Likewise, "survival of the fittest" hints at a process of elimination but it has come to be interpreted positively. When we gloss over the fact that evolution doesn't explain the origins of life, we lean positive, almost as though natural selection was indeed some new physical law that began, at some point in natural history, to promote life.

Indeed, whenever we talk of physical laws we lean positive. Laws can be imposed to either eliminate or produce behaviors, but in the imposition, we subconsciously assume a positive force, something promoting events.

EMERGENCE AS POSSIBLE DYNAMIC PATHS DRAINED AWAY

These days, research to solve the mystery of purpose is treated as a focus within the field called *emergence*, a field that addresses how there can be new things under the sun.

The term *emergence* originally meant to extract something from its merged state. It later came to mean something *revealed* that hadn't been visible before, the subjective observer having overlooked it.

In keeping with the second meaning, maybe selves and aims have always existed but have simply gone undetected. Sure, nothing in the universe's first ten billion years appears to be a self with aims, but that's because we just haven't yet detected that they were here all along.

This interpretation of the term *emergence* is consistent with *panpsychism*, the argument that even subatomic particles have the capacity to aim that somehow aggregates into recognizable selves. By such a reckoning, selves and aims were always present, just undiscovered.

Here, our argument is that selves and aims haven't always existed. They truly are something new under the sun. Still, their existence was always possible, as is evident from the way that their behavior doesn't violate any physical laws. There's nothing new about the possibility of them existing, but there is something new about the likelihood of them existing.

Here we'll return the term *emergence* to its original meaning, applied not to things but to dynamic paths. Selves did not always exist, merged in the universe like carrots merged within a bowl of soup and waiting to be discovered when ladled out. Rather, among the myriad dynamics always possible in the universe, the very rare dynamic paths that make selves possible did exist as a slim possibility, merged with all the other possible dynamic paths. They emerge through a negative process of elimination, a local constraining away of alternative dynamic paths.

Selves and aims were not always present, waiting to be discovered. Nor were they some new addition to possible dynamics. Rather, they are a possibility that becomes more likely by a chance process of elimination that constrains away alternative dynamic paths. Explaining the origin of selves requires that we identify the chance process of elimination that locally constrained away the alternative dynamic paths that don't result in selves with aims.

SPELUNKING FOR SELVES

Unpack any self's body and all you are going to find in it is cause-and-effect material dynamics operating in full nonmagical compliance with the laws of nature. But not just any dynamics. The paths that are not present explain the paths that are. *Prevented dynamics account for presented dynamics.*

Materialists have been like miners who never take off their headlamps. Everywhere they turn, they see material things behaving in accordance with natural laws. They conclude that we have examined everything and it's all just material dynamics. Metaphorically, they don't see the peripheral darkness—the possible dynamic paths that are not present and therefore cannot be seen.

Ashby's "what might have happened but did not" has no mass, volume, charge, or parts, none of the attributes we associate with material objects. It cannot be analyzed reductively. That is, you can't take apart what didn't happen to see what materials it is made of. But still, what doesn't happen makes a difference.

As Ashby states, "while, in the past, biologists tended to think of organization as something extra, something added to the essential variables, modern theory based on the logic of communication [Shannon's theory] regards organization as a restriction or constraint."[2]

Thus, though we shall be looking for the key where it's lightest and not in the dark realm of quantum mechanics, there is a crucial kind of dark we have overlooked that here we will highlight: the absence of once-possible paths that have somehow become constrained toward lower probability.

Darwin saw the light by examining this dark. He saw the lineages that *didn't* survive as crucial to explaining those that did. The traits that were eliminated in the lineages that didn't last mattered. If this variety hadn't been produced and then culled by being outreproduced, the lineages that persisted wouldn't have fared as well as they did in their environment.

In this, he was a bit like Sherlock Holmes in "Silver Blaze," who determined that the murder was an inside job since the dog didn't bark, as he would have if a stranger had been the culprit. Absences can sometimes be evidence.

WHAT'S PRESENTED FROM WHAT'S PREVENTED

Rather than simply focusing on what promotes or produces selves, we'll be looking for what reduces the chances of selves dying. We'll be concentrating on how selves negatively constrain and therefore prevent their own death. "Purposes," "selves," and "aims" state positively what is achieved, but for us, the key to explaining them will be constraints that prevent them from ending, as though the most fundamental purpose is prevention of a loss of our ability to have purposes. The most fundamental aim is preventing the loss of our ability to have aims, and the must fundamental quality of selves is the self-regenerative prevention of the end of their selfhood.

The emergence of selves and aims, and therefore the solution to the mystery of purpose, will thus rely on a double-negative kind of reasoning, not what says yes to life but what says no to dying. From this perspective, life emerges by subtraction—negative processes of elimination and not addition or philosophical abstraction.

The marvelous things that selves can do result from dynamic possibilities that have been prevented from occurring. There's nothing about what you do that wasn't always physically possible. It was just extremely unlikely before. Through emergent constraint, your unlikely mental and physiological capacities have become increasingly likely. Reduced variety means more similarity, which is why you are so similar from day to day and why lineages of organisms have maintained dynamic continuity over eons, despite the second law of thermodynamics.

Deacon argues that a self is neither a ghost nor a ghost-made machine—just dynamics playing out in a natural context. That doesn't leave us agape at some unsolvable mystery. An alternative to both ghost and machine has been long overlooked given our materialist assumptions. Selves and aims are the dynamic paths that remain after other dynamic possibilities have somehow become less likely. How that happens is the mystery we can solve now that we know to look at what's eliminated to make selves, aims, and purposes possible.

We'll be focusing next on how constraints that eliminate dynamic pathways work. And we'll start with the most fundamental physical tendency, the second law tendency for unconstrained dynamics to become disorganized, or more specifically irregular.

16

SECOND LAW IRREGULARITY

WHICH IS MORE COMPLEX: A FROG OR A FROG SMOOTHIE?

Life is *complex*, but oddly not in the primary technical sense of the term. In most technical treatments, a more complex thing takes longer to describe than a simpler thing.[1] For example, which is more complex: the series 00000000 or the series 01110100? Intuition and technical definitions converge in identifying the second series as more complex since there's no obvious pattern. The first is simple, just ten zeros. The second would take longer to describe since it's irregular.

Irregularity is complex; regularity is simple. The term *regular* comes from the Latin *regula*, meaning "moving in a straight line," from which we get the word *ruler*, marked off in regular units, one after the other, like our unvarying sequence of zeros.

A regular customer invariably goes to his restaurant and invariably orders "the usual," his regular dish. The regular customer's eating habits are less complex than the eating habits of the irregular customer.

Now suppose either customer orders frog, which is served two ways, whole or blended. Which is more complex, the whole frog or the frog smoothie? This is where intuition and the technical definition of complexity diverge.

The whole frog, with its various organs, textures, and contours, appears more complex than a nice, homogeneous frog smoothie, but the whole frog is actually less complex. The whole frog has more regularized regions, for example, regular sequences of bone or skin molecules arrayed like the series of zeros.

Though the whole cup of frog smoothie appears to be one smooth substance, at the molecular level it is more complex—less regularized than the whole frog. Examining neighboring molecules in the smoothie, one finds an irregular, heterogeneous molecular distribution—skin and bone molecules randomly distributed, configured complexly anywhere within the smoothie tumbler. It would take a lot to describe the smoothie's heterogeneous distribution of molecules. The smoothie is blended toward irregular distribution. The whole frog is segregated into highly regularized distributions.

To solve the mystery of purpose, we must explain the whole frog's segregated regularities and their presence despite the most unrelenting and all-pervasive blender of them all.

That blender is the second law of thermodynamics, the tendency for regularities to become irregular over time, in other words, for simple order to become complexly disordered, for the nicely organized to become disorganized or all mixed up. Counterintuitively, the second law is a blender that needs no power supply. Instead, as we'll see, it is a power source.

Here we'll take a brief intuitive tour of the second law and its implications for solving the mystery of purpose. Its implications are many. The second law tendency toward irregularity is what selves must resist and overcome if they are to survive and reproduce.

But the second law is also the source of energy and work, including the work that selves do to resist and overcome the second law. Thus, through their ability to aim or focus energy into work, selves use the second law against itself.

SAME DIFFERENCE

Pour cold milk into hot coffee and you get warm milky coffee. Open the valve on a tank of high-pressure gas in a low-pressure room and the pressures equalize. Connect the terminals of a battery and you will eventually get a dead battery. Open a perfume bottle in a room and soon the perfume molecules are distributed randomly throughout the room.

In each case, once-segregated regularities become irregular. Start with regularized or homogenous concentrations segregated apart from one another and allow them to interact and they naturally become irregular distributions. Sorted concentrations become mixed up, segregations desegregate, organization becomes disorganized, order becomes random, distinct homogeneities blend to become heterogeneous.

Scientists measure irregularity in terms of *entropy*. At maximum entropy dynamics are maximally irregular—maximally complex. The second law is the conventional name for the tendency for dynamics—large populations of elements interacting—to fall passively toward maximum entropy, irregularity, or complexity. Selves defy this tendency. As we'll see shortly, defying it sustainably is the defining feature of selves and aims.

To focus on one example of the second law, before we poured the milk into the coffee, the milk was a regularized collection of similar, relatively slow-moving milk molecules like our series of zeros. The coffee was likewise a regularized collection of similar, relatively fast-moving coffee molecules, like a regular series of ones.

Once the milk and coffee interact, they spontaneously mix, becoming an irregular collection of slow and fast milk and coffee molecules in an unpredictable, random distribution, like an irregular series of zeros and ones. It would take a lot to describe its irregular configuration. Like the cup of frog smoothie, the cup of milky coffee looks uniform and therefore simple, but at the molecular scale, it's complex.

We recognize irregularity by our inability to find regularities within the distribution of molecules. Knowing the nature of one molecule tells you nothing about the nature of neighboring molecules. They have become decorrelated or desegregated, like a desegregated neighborhood where knowing the nature of one neighbor tells you nothing about the nature of another.

The global uniformity (the appearance of homogeneity in the smoothie or milky coffee) that results from underlying irregularity is nicely captured by the colloquial oxymoron "same difference." Like the oxymoron "jumbo shrimp," it is not a contradiction if you switch scales of analysis halfway through the oxymoron. Shrimp is the macro class; jumbo is the micro within the larger class of shrimp. "Same" is the macro uniformity; "difference" is the micro complexity within the macro similarity. At maximum entropy, distributions are uniformly random, the same difference everywhere.

Scientists count the second law as one of the most, if not *the* most, reliable tendency in the universe. Irregularity happens. Paradoxically, the most regular thing in the universe is this tendency toward irregularity.

IRREGULARITY HAPPENS

Nothing imposes the second law. It is simply a product of probability when interaction is allowed. If you have two trays of coins, one heads, the other tails, and you toss them together, they're very unlikely to land in any kind of regular distribution of heads and tails. There are vastly more possible irregular distributions than regular distributions. And the more coins, the more likely an irregular distribution is.

If different collections of regularized molecules are free to interact, they'll fall toward irregular configurations simply because there are vastly more of them. And as the molecules continue to interact, they will pass from one irregular configuration to another, again because there are so many more of them.

Irregularity does not result from either a process of production or elimination. It's just all possibilities equally presented. With vastly more irregular possibilities than regular ones, the irregular possibilities are the most likely outcome by a lot.

Given the highly skewed natural distribution of probabilities with vastly more irregular than regular configurations possible, it takes work to make and keep things regular, to sort the naturally occurring irregularity into segregated regularized order. To *regularize* is a verb, an action word. *Irregularize* is barely a word, and that's good for our purposes because the second law is not an action. The second law is not an imposed law. It takes no work to generate disorder.

Sure, we can "irregularize" things, for example, working to get that frog blended quickly for an impatient customer. But if we let the dead frog be, its molecular distribution will become irregular anyway, as does the molecular distribution at death for all of us. Being alive requires maintaining the body's segregated regularities. When we stop doing that, we're dead.

Death is the irregularity that happens spontaneously when the frog's work to maintain its regularities stops. Death is like becoming the smoothie. Dying is easy. Staying alive takes aimed work, the work we must explain.

Work, of course, takes energy, defined as the potential to do work. About energy, too, people's intuitions diverge from technical understanding. We intuit that energy is a substance or force that can be used up, but energy is neither created nor destroyed. When something "loses" energy, the energy hasn't actually been lost but, in effect, diluted, equalized, diffused, or desegregated—so irregularly distributed that it can no longer do work.

ENERGY IS IRREGULARITY HAPPENING

Slip a frozen pizza into a hot oven and the pizza cooks. Ignite gasoline in an engine cylinder and the car moves. Connect battery terminals with a motor in between and the rotor spins. In each of these cases, segregated energetic regularities are allowed to interact. Given the second law, the segregated regularities fall toward irregular distributions. That is what energy is: regularities falling toward irregularity given the second law.

With the pizza, at first, there's regularity, molecules consistently slow within it compared to the consistently fast-moving air molecules in the oven, which, for simplicity's sake, we'll assume is well-insulated, preheated, and turned off. Simplifying, if we translate molecular speeds into zeroes and ones, then, prior to cooking, the molecules in the frozen pizza are 00000 and those in the oven air are 11111. By putting the pizza in the oven, we allow the molecules to interact. Soon molecular speeds fall into irregular sequences like 0110100100.

The pizza and oven temperatures equalize globally, but at the molecular scale, temperature equalization is the result of molecular speeds becoming distributed more irregularly, with fast and slow molecules in any location throughout the pizza and oven. Temperature, of course, is a function of molecular speeds. Again, *same difference*: equalized temperature at the global scale and irregular speeds at the molecular scale.

During cooking, the molecules interact. Slower molecules accelerate when hit by faster molecules and vice versa. At the macro scale, we see temperature equalization between oven and pizza that, contrary to intuition, isn't a result of the molecular speeds all becoming the same but rather randomizing, the molecular speeds becoming irregularly distributed throughout the combination of oven and pizza.

Words like *energy* and *heat* are nouns. As such, they can give us the false impression that they are things or at least forces. They aren't. Rather they're the second law playing out, dynamics becoming globally homogenous (for example, temperature equalizing between pizza and oven) by becoming, at the molecular scale, heterogeneous (for example, molecular speeds becoming irregular).

What we call heating is simply the interaction between formerly segregated, regularized fast and slow molecules becoming irregularly distributed through interaction. And cooling is the same, as when we cool our coffee by allowing the slow-moving milk molecules to interact and fall toward random distribution with the fast-moving coffee molecules.

The same applies to all forms of energy. An explosion is not the release of a substance or force called energy or pressure. It's the equalization of formerly segregated regularizations. What we call energy would more aptly be called an energetic gradient or difference equalizing due to the second law.

There need not even be energetic gradients involved for irregularity to happen. When we uncork perfume and the molecules distribute randomly throughout the surrounding air, there's no energetic gradient. Still, there's the tendency toward irregularity. Though the second law originates and is mostly applied to energetic differences equalizing, I will broaden its application. Given the second law, segregated regularities fall toward irregularity regardless of whether the segregated regularities are energetically different.

WORK IS THE SECOND LAW CONSTRAINED

Energy is the potential to do work—potential that doesn't manifest as work unless there's an interaction between two formerly segregated and therefore different regularities. Think of the difference between throwing a punch at nothing or something. With nothing in its way, the fist does no work. With something in its way, work happens. Work is the equal and opposite reaction between divergent dynamic tendencies interacting with each other.

Let's picture this with the pizza and oven and at three scales: individual molecules interacting, interaction throughout either the pizza or the oven's dynamics separately, and interaction between two formerly isolated, now-interacting dynamics.

Let's start with the oven, a hot insulated box of air molecules interacting with one another. Through interaction, the air molecules are doing work, accelerating and decelerating one another. The resulting molecular speeds are irregular or heterogonous.

Before we slip the pizza into the oven, the oven overall has a homogenous temperature, same difference from one corner to another within the oven. Likewise the frozen pizza, sitting in the freezer prior to cooking, has a homogeneous temperature. Though the molecules within it are interacting, the pizza is roughly the same frozen temperature throughout.

Once we slip the pizza into the oven, the difference between the pizza and oven temperature interact. Given their temperature differences and the second law, the formerly segregated molecules in the pizza and oven interact and work upon one another.

Work and therefore energy—the potential to do work—are relational features playing out at different scales of analysis. The molecules in the frozen pizza and hot oven do work on one another because there are energetic differences between them. The work is exclusively molecule on molecule, but because the molecular speeds are regularized independently in the cold pizza and hot oven, when the two interact, there's a direction of change over time. The pizza gets hotter; the oven (for simplicity a preheated box) gets colder.

Is the oven filled with a stuff called energy or heat? If you tossed it into a volcano, the oven, being cooler, would get hotter. Heat, like any kind of energy, is a function of the relationship between one thing and another. Energy is actually a difference between things that could interact. If they interact, they do work on one another. A stationary wall in isolation doesn't contain the energy to do work on things, but crash into it and you'll find that it has plenty of capacity to do work, but only on things with different dynamic tendencies.

SPRING-LOADING THE SECOND LAW

With a battery and motor, the energetic character that becomes irregular isn't molecular speed but charge, an atom or molecule's ability to take or give up electrons. The simplest charged battery is two isolated regularized collections of molecules, one collection negative (– – – – – –), or able to give up electrons, the other collection positive (+ + + + +), able to accept electrons.

Bridge the battery terminals and a dynamic path opens. The once-insulated regularizations – – – – – – and + + + + + become irregular sequences (– + – – + + + – + –, or any other random variation). If we connect an electric motor in the path between the terminals, the dynamic path from regularized to irregular does work, turning the motor's rotor until the battery dies, the charges having reached maximum entropy, in other words, maximum irregularity.

A discharged battery exhibits no directional change. The positive and negative charges haven't been lost, just irregularly distributed even as the molecules continue exchanging electrons, moving from one irregular distribution to another.

Charging a battery takes work. When you charge a battery, you're basically reregularizing those positive and negative charges into segregated cells. It doesn't take work for the battery to discharge any more than it takes work for a ball to roll downhill, or a pizza to heat up in the hot oven.

It takes work to regularize things. Given the second law, irregularizations are effortlessly inevitable. Paradoxically, when we think of something as doing work—for example, the hot oven working on the pizza, or the battery turning the motor—there's no macro work being done, just the micro molecule-on-molecule work of molecules interacting. The macro work was done prior, in effect spring-loading the second law tendency. It takes work to produce segregated regularities that will then desegregate toward irregularity if allowed to interact.

And notice that the work it takes to spring-load such regularized differences is not just any work. It takes highly constrained work to heat an oven and freeze a pizza or to get all those negative and positive charges segregated in a charged battery.

CONSTRAINED PATHS

If you went outside in your underwear in the dead of winter, your warm-blooded body would interact with the cold air. You would lose heat by many paths over the whole surface of your body. If, instead, you were wearing insulating skiwear, the paths for heat exchange would be constrained, for example, off of your face faster than elsewhere.

Don't try this at home but in your imagination: Hang a can of gasoline from a tree with a fuse hanging out it. Ignite the fuse and watch the can explode omnidirectionally—multiple paths for the second law tendency toward irregularity to play out. When, instead, gasoline explodes in a car engine, the hard metal cylinder constrains the explosion's possible paths. The cylinder imposes constraints that limit the paths by which the second law tendency toward irregularity occurs. Constraints channel energy into regularized work.

Controlling, harnessing, directing, channeling—we use such terms to talk about how we get work done. Though we pay attention to the work thereby achieved, these terms all imply our negative approach, a process of eliminating or constraining possible paths down to those that enable us to achieve our aims.

Engineering anything is largely a matter of constraining energetic gradients so that they achieve the work we want done. Whether designing giant dams or smart phones, roadways or triple heart bypasses, engineers act as entropy-wranglers directing the second law tendency toward irregularity, corralling and channeling energy gradients into functional dynamic paths by means of constraint. Functional machines don't generate their own energy. They channel the equalization of energetic differences down constrained paths.

PROCESSES OF ELIMINATION, PATHS OF LEAST RESISTANCE

A closed valve stops water flow. A switch in the off position stops a battery current from flowing through a wire. An explosion occurring within a solid container prevents material from flying out of the container.

In each of these cases, paths toward equalization are eliminated because the resistance is too great to overcome.

Water pipes or electrical insulation blocks alternative paths for currents of equalization. The only remaining paths are at their ends, water flowing out the pipe or current flowing out the end of the wire into whatever it's connected to. A closed valve or an electrical switch in the off position constrains these remaining paths.

Constraint is relative, nicely captured by the idea of paths of least resistance. If water pressure is low, then the closed valve stops water flow completely, but if the pressure is high enough, it can break the valve or the pipe, whichever imposes the least resistance.

A valve on a water pipe imposes a solid constraint on water flow. Not all imposed constraints are solids. Distance can also be a constraint. To stop an electrical current we can either impose a solid piece of insulation or open a gap in the conductor. Both will isolate regularities so that they can't interact.

The second law tendency toward irregularity is so reliable and universal that it is how time becomes intuitive to us. We mark time by the second law tendency toward irregularity. We wind up clocks, creating pressure differences that release over time. We mark time with electric clocks that, like motors, run energetic differences toward irregularity. More generally, we know the direction of time by the second law tendency toward irregularity. We expect eggs to break; we don't expect broken eggs to spontaneously reform into whole, highly regularized eggs. If we saw an egg spontaneously reregularize, we'd think time was going backward.

Hence the mystery of purpose, because while broken eggs don't spontaneously reregularize, birds regenerate regularity to molecular configurations that otherwise would be randomly distributed all over the place. Organizing an egg from disorganized molecules takes work. This is the peculiar kind of work we must explain from its origin, the work done by selves to generate and maintain their own regularity.

17

EMERGENT REGULARIZATION

IMPOSED VS. EMERGENT CONSTRAINTS

Cylinder walls, insulated wire, and gullies—these exemplify the way imposed constraints channel the second law tendency toward irregularity into regularized paths, energy currents flowing in one general direction instead of many.

Sometimes we imagine that the regularized currents in selves are the products of imposed constraints too, as though we are "hard-wired" machines. There are certainly some imposed constraints within our bodies. Myelin imposes constraints on neuronal electrical currents. Veins and arteries impose constraints on blood flow. The digestive tract imposes constraints on the flow of food nutrients. The pulmonary system imposes constraint on airflow.

But these are not imposed from outside like a girdle. Life's organization isn't imposed by protoplasm flowing into imposed molds, natural or engineered. Imposed constraints do not explain all regularizations.

This is why *self-organization* has become of fundamental interest for many origin-of-life researchers. Self-organization is regularization that originates *throughout dynamics* rather than being imposed from outside the dynamics. *Throughout* may seem an awkward word choice, but as we will see, it will prove necessary to avoid common confusions about self-organization.

To illustrate self-organization, consider a river flowing downstream, constrained and regularized by a smooth channel—an imposed constraint.

Water flows in at the inlet and out at the outlet at roughly the same rate. The flow is laminar (smooth), regularized by the imposed channel such that instead of many paths out there's one general path.

If we place an obstacle somewhere in the channel, partially constraining the outflow, we get turbulence, currents working against currents. In this work, paths will constrain one another, altering the relative likelihood of some currents compared to others. The irregular turbulence will, under certain conditions, self-organize, resulting in a whirlpool or eddy, a regularized spiral current.

For an intuitive sense of how this happens, imagine a disorganized crowd of people streaming turbulently from all directions into a foyer but all having to exit through one door. People are, of course, selves with aims, but let's ignore that for the moment. Assume that they're all lost in thought about other things and are being shuffled toward the only exit simply because other people behind them are pressing them forward.

At first there's no regularity to their exit pattern. People take whichever paths impose the least resistance. But traffic currents begin to form. Since people are entering the foyer from many doors, some of the currents cross, resulting in impasses that the people begin to circumnavigate, shifting into faster currents. Some currents amplify; others dampen. With time, the exit traffic simplifies down paths of least resistance. For some people, the path will be the shortest, a straight shot from their foyer entrance to the only exit. For other people, the exit paths will be more circuitous.

The resulting paths are the product of constraint, though, importantly, not imposed constraint. There are no walls or guiderails imposed on the exiting crowd forcing it to take their constrained paths. Rather, the currents *emerge*, the product of something like the constraint we find in traffic congestion, like the dynamic constraint you experience when you're threading your way through congested traffic, and how in your threading, you impose constraints on others threading their way through it.

NET-WORK

What then accounts for a whirlpool's spiral regularized current? Imagine cars at an unmarked intersection, with traffic entering from many angles.

Gridlock would cause cars to slow to a standstill or shunt them into whatever traffic patterns flow around the gridlock. In time, the cars are likely to simplify into a circular current, a roundabout, the circular form that traffic engineers have discovered is the most efficient way to keep cars moving through an intersection congested by cars entering from many angles.

A whirlpool is like a roundabout formed as turbulent water currents simplify into water-flow paths of least resistance by a process of elimination, some paths becoming less likely, leaving other paths more likely. Once the flows enter paths of less resistance, they will not spontaneously jump onto paths of greater resistance any more than a ball rolling downhill will spontaneously roll back up.

A whirlpool is a dynamic regularity, water flow in a spiral direction. We can think of a dynamic flow as a particular kind of network. The term *network* often means a complex of roads, or a group or system of interconnected people or things. If we concentrate on its meaning as a system of interconnected things—in the whirlpool's case, water molecules—the interconnectedness is in the molecules' capacity to do work on one another, moving one another along. Networks are the paths in which work is likely to occur.

The term *network* originated in the 1550s to mean a netlike arrangement of threads or wires. But the word *net* has two origins that converge to its current meaning. Its older meaning is "knot," as in a knotted fishnet. But net also derives from the same root as *neat*, which is how it comes to mean "remaining after deductions," as in net worth. Once you run things through netting, they become neater.

Fishnets are materially imposed constraints that allow only some material to pass through, but as the whirlpool example illustrates, not all constraints are imposed. Some are emergent, resulting from work throughout a system. These emergent constraints filter out—or "net out"—other possible paths. People exiting a foyer or water molecules exiting a bathtub become their own netting from which neater, simpler regularized paths are netted.

A dynamic regularity like a whirlpool can thus be thought of as a network of work paths that water can take after the turbulent paths that don't work are netted out, in effect by impasses. This eccentric use of the term *net-work* suggests a way to think of how dynamic form can emerge.

WHAT DOESN'T CAUSE THE SPIRAL

Let's be very clear about what causes a whirlpool's spiral net-work. It isn't caused by water flowing, or by the obstruction. These imposed constraints are necessary conditions for the emergence of the whirlpool but they don't impose the spiral regularization.

Researchers sometimes refer to the spiral as an "attractor state," which could give the false impression that water is being pulled into its spiral form as though by a magnet. This false impression would be consistent with our materialist intuitions that all change is caused by pushes or pulls. But no, there's no attractor within the water pulling water flow into the whirlpool's regularized spiral path.

Researchers also often describe self-organization of something like a whirlpool as *top-down* or *downward causality*, a term coined in 1974 by the same Donald Campbell who described evolution as "blind variation with selective retention." Top-down causality implies that there is a top level—a whole system—causing, as if pushing, the turbulent water into its spiral path. As with talk of attractors, talk of top-down causality conforms to our push-pull intuitions about all causality. But no, there's no top-level whole that pushes water flow into a spiral current.

How could there be? In water flow, there's only molecular interaction, molecule-on-molecule, cause-and-effect work. This work doesn't somehow conspire to elect some higher source of governance over the molecular behavior. As Jaegwon Kim, the philosopher famous for his critique of top-down causality declares, "There are no causal powers that magically emerge at a higher level and of which there is no accounting in terms of lower-level properties and their causal powers."[1]

The whirlpool's spiral is not imposed from outside, not pulled from inside, and not a bottom level that produces a top level that pushes the bottom level into conformity. So what explains the spiral? It results from molecule-on-molecule work *throughout* the turbulence.

The whirlpool is a product of constrained interaction throughout dynamics. It is not a product of constraints imposed upon the dynamics from other sources. The whirlpool is not engineered of course, and yet it is a constrained regularized current. As such it is a source of emergent regularity, an energy gradient capable of doing sustained work. A bathtub whirlpool can't do much work, but other vortices can. Tornadoes can cause houses, cows, cars, Dorothy, and Toto to spiral skyward.

THE MISNOMER *SELF-ORGANIZATION*

Self-organization, as exemplified by whirlpools, is an important source of novel constraints, overlooked for centuries and a crucial stepping-stone to understanding how selves and aims emerge. But it's poorly named in several respects.

First, *organization* is a vague term. It can imply either a preferred configuration or any configuration at all. One can even say, "Due to the second law, most dynamics have an irregular organization."

What we see in the whirlpool is more accurately described as regularization. *Regularization* is also more accurate because it suggests a question of degree in a way that organization does not. *Organize*, as a verb, implies a process of achieving organization, a single, static, final state. In contrast, *regularize*, as a verb, implies a process of becoming gradually more regular, which, unlike organization, suggests not a single, static, final state, but any of a variety of states in which irregularity is reduced by some degree or other. The whirlpool doesn't reach a final, static, organized state. It regularizes to some degree.

We could thus more accurately describe self-organization as *self-regularization*. Ashby recognized the problems with his term *self-organization*, and at times referred to it as *self-simplification*, a term that points to regularization, since regularity is simpler than complex irregularity. Still, simplicity refers to our interpretation of it more than its distinguishing physical feature.

And the prefix *self-* is ambiguous too. It often means *not imposed from the outside*, as in *self-starting*, but it can also suggest the existence of a real self, as in self-motivated. And sometimes the prefix *self-* simply means internal to some well-defined system, as in *self-contained*. In any of these uses, it implies that an inside and an outside are clearly defined. A whirlpool is not clearly defined. It is merely dynamics in a general locus of individual molecules interacting.

This analysis of the use of *self-* as a prefix might seem nitpicky if it weren't that so many dynamic systems theorists suggest that self-organization can completely explain how selves emerge. The suggestion that selves emerge merely through self-organization is an example of *thresholdism*, the idea that once interactions cross a threshold a self just happens.

Given the problematic implications of the term *self-organization*, I will refer to self-organization as *emergent regularization*. A whirlpool's spiral current is an emergent regularity—molecules moving down an emergently constrained path, all in more or less the same spiral direction like regular customers going the same way to the same restaurant.

The whirlpool's spiral path is not the product of imposed constraint. There's no external channel limiting water flow to the spiral path. Rather, the path is the result of emergent constraint, alternative currents getting in one another's way. Self-organization, from here on called emergent regularization, is one of two kinds of emergent constraint. Self-regularization is a necessary but insufficient condition for the second kind of emergent constraint, which I'll call *self-regeneration* and detail shortly, since it is what Deacon argues emerges as selves and aims.

EMERGENT REGULARIZATION

Both the spiral whirlpool and the second law tendency toward irregularity are merely the result of molecule-on-molecule work throughout dynamics. But notice that the second law and the spiral dynamics have opposite consequences. The second law tendency is toward irregularity but the whirlpool's tendency is toward regularity.

This is what makes emergent regularization so interesting. When we start an experiment with water flowing around an obstruction, at first it tends toward the turbulent irregularity we expect, given the second law. But then, without a change to imposed constraints, it begins to regularize into a whirlpool. In general, we don't expect results to change midway through an experiment. For example, Galileo would not expect a ball rolling down a ramp to suddenly change directions, unless some new constraint was imposed that does work on the ball. With the whirlpool no new constraints were imposed and still there's a switchback from expected irregularity toward regularization.

The whirlpool's tendency toward regularity is entirely predictable given the laws of physics. Our materialist deterministic intuitions incline us to search for what caused it, what object or force external to the dynamics was imposed causing them to become regularized.

Nothing was added. Rather, possible dynamic paths were subtracted by interaction throughout turbulent currents. This is what justifies calling the resulting regularization *emergent*—emergent in its original sense—possible dynamic paths that were always present becoming more likely because other paths became less likely as a result of impasses or gridlock, currents checking other currents. Paths that are emergently prevented explain the paths presented.

To be clear, the emergent regularizing phase is the gradual increase in regularity. Once it arrives at regularity, it is no longer emergently regularizing but is merely regularized. A whirlpool can become stable, not staying in a single regularized static state but visiting a limited range of regularized states.

Emergent regularization yields a variety of natural physical forms, including Bénard cells (regular hexagon patterns in heated oil), autocatalysis (chemical chain reactions generating a regularized concentration of certain molecular types), and crystal formation, all of which we'll explore briefly later. In each case, there's a direction of change toward regularization. But first, we need to explore what emergent regularization can and can't yield. It can't yield selves. Selves emerge as a result of the second kind of emergent constraint, *emergent self-regeneration*.

18

EMERGENT REGULARIZATION VS. EMERGENT SELF-REGENERATION

A FOOTING, NOT THE WHOLE BRIDGE

Second law irregularity and emergent regularization are tendencies that yield opposite dynamics. Does emergent regularization therefore defy the second law? Hardly. The whirlpool is a local regularization that actually accelerates the global tendency toward irregularity. A whirlpool is, after all, a path of least resistance for water flow.

Water drains out of bathtubs faster with a whirlpool at the drain than with turbulence. The whirlpool persists precisely because there's no faster alternative path for water to escape. For a whirlpool to spontaneously become turbulent would be like water flowing downhill suddenly U-turning to go uphill. It just doesn't happen, and not because anything about whirlpools aims to achieve efficiency. Water falls efficiently, and once it falls into a more efficient path, water won't climb back to a less efficient path.

Emergent regularization in whatever medium is a temporary local regularization that accelerates the second law tendency toward global irregularity. If a whirlpool were a self, it would be a short-lived one. Emergent regularization is thus a runaway process that eventually ends, succumbing to irregularity by the path of least resistance.

A whirlpool is just the second law, with constraints emerging throughout the dynamics yielding a local, temporary increase in regularity, regularity that actually speeds up the global tendency toward irregularity in comparison to turbulence. Order is the fastest path toward disorder. Regularization is the fastest path toward irregularity or maximum entropy.

Some researchers argue that like whirlpools, selves are just the universe's way to produce irregularity faster than it would otherwise occur. For example, dynamic systems theorists Rod Swenson and Michael Turvey argue that "The world is in the order production business, including the business of producing living things and their perception and action capacities, because order produces entropy faster than disorder."[1]

This intuition is understandable given current social trends, such as burning through millions of years of accumulated fossil fuels in a few generations. In large part because of our ability to exploit fossil fuels, humans are undergoing a rapid population explosion, extracting and consuming ever more resources, transforming them into more people who are extracting and consuming more resources still.

Thus, these researchers are onto something about the human population overall. A population explosion is actually an example of emergent regularization occurring throughout the medium of selves. Think of emergent regularization as a compounding process that makes something more regular but in the process depletes the resource it depends upon. Increasingly, we are becoming regular customers for fossil fuels. The more we use, the more we are likely to use, until we run out, perhaps even at risk to human survival.

But these researchers are wrong to argue that selves in general are emergent regularizing dynamics that speed up resource depletion, and fossil fuels are one clue why. Fossil fuels are an enormous repository of highly regularized chemicals left behind by selves—mostly unicellular algae selves that produced regular dynamic forms that persist over their lifetimes. At death, some of that regularized form is retained as deposits underground. If life were merely the universe's ways of accelerating the tendency toward irregularity, there would be no such accumulated repositories.

Humanity's emergent regularizing growth spurt over the past two hundred years was made possible by regularities accumulated as fossil fuel over prior millions of years. If selves were as self-undermining as whirlpools, they would not leave behind fuel that could be burned today. They would dissipate energy as efficiently as possible.

NEGENTROPY

Toward the end of his career, the Nobel Prize–winning physicist Erwin Schrödinger wrote a short book called *What Is Life?* in which he sought the underappreciated aspects of physics that make life possible. He argued that "living matter, while not eluding the 'laws of physics' as established up to date, is likely to involve 'other laws of physics' hitherto unknown, which, however, once they have been revealed, will form just as integral a part of this science as the former."[2]

Schrödinger made the case for two distinct features of living systems. One was a hint that contributed to the discovery of DNA. The other, equally important but underappreciated, is that life defies the second law tendency toward irregularity.

Recall that maximum entropy is maximum irregularity. Schrödinger coined the term *negative entropy* for what life produces—sustained regularization despite the universal second law tendency toward irregularity. Later, physicist Léon Brillouin shortened the term to *negentropy*, a name that stuck. Emergent regularization produces local, temporary negentropy while maximizing global entropy.

Still, the negentropy achieved by emergent regularization is decidedly different from that achieved by selves, which is regularity maintained over lifetimes, generations, and eons.

Most obviously, there are no lineages of emergent regularization dynamics. Whirlpools don't spawn baby whirlpools. In contrast, your body has regularities inherited from your earliest ancestors, regularities that you maintain and can pass on to future generations.

Most importantly, emergent regularization does no work to maintain its own regularities. Turn off the water or remove the obstruction and the whirlpool disappears. Nothing about its regularities puts up any resistance to its termination. The whirlpool's existence is at the mercy of its imposed constraints, for example, the obstruction and the dynamic water flow. If anything, it eliminates those conditions since it speeds up the draining off of the water flow that makes it possible. In contrast, selves put up resistance to their eventual dissipation, and our goal here is to explain how selves do that.

SELF-REGENERATION

We thus must distinguish between emergent regularization that, for example, results in whirlpools and what I will call *emergent self-regeneration.*[3] This distinction will prove pivotal to solving the mystery of purpose and I make it in direct refutation to thresholdist arguments that self-organization explains or explains away selves and aims.

Earlier, I replaced the term *self-organization* with *emergent regularization*, in part because the prefix *self-* is ambiguous and hints at a self where there isn't one. Now, I'll employ the prefix *self-* in self-regeneration, arguing that, with self-regeneration, a true self exists, an argument I will have to justify later.

Self-regeneration is characterized by three fundamental capacities, functional traits that all selves possess and are not present in any emergent regularization dynamics. All three are ways that selves defy the second law tendency toward irregularity, thereby generating Schrödinger's sustained negentropy:

1. Self-Repair: To overcome second law irregularity, selves constrain or aim energy into work to regenerate their regularities. This aim manifests in a self's capacity to acquire energy and resources, which it converts into work to repair or heal, to regenerate regularities that would otherwise fail through second law degeneration.
2. Self-Protection: Among the features that selves regenerate are protective barriers, constraints that protect and segregate the self's regularities against second law degeneration, for example, cell membranes, seed hulls and shells, skin, and exoskeleton. Repair and protection complement each other. Repair regenerates the regularities that protection locks down, protected from second law tendency.
3. Self-Reproduction: Selves proliferate selves, dynamics that inherit the three features of self-regeneration. They do so with evolvable variety—Darwin's heredity and variation.

One might assume that reproduction is the self's first priority, but given the second law tendency toward irregularity, the self's first priority is repair and protection, regenerating what the second law degenerates and protecting against that degeneration. A self that has lost the capacity to repair and protect its regularities is dead and therefore can no longer reproduce.

SELECTIVE INTERACTION

The three features of self-regeneration impose opposite demands. Self-repair and self-reproduction require energy and resources and thus depend upon energetic and material interaction with their environments.

In contrast, self-protection requires isolation and insulation, the prevention of interaction with the self's environment. Selves must stay closed to energetic gradients that would render their regularities irregular.

Paradoxically, to remain an independent self, a self cannot be entirely independent. Selves are islands, but islands that selectively import and export. They must be open to the energy and resources that make self-regeneration and self-reproduction possible, but closed in ways that make self-protection possible. Deacon calls this "the paradox of autonomy," the self's need to be both open to the right interactions and closed to the wrong interactions.

All known organisms have an evolved capacity for selective interaction. Cells have their evolved semipermeable membranes, plants and animals have their evolved pores, and animals have their evolved orifices. The paradox of autonomy is easiest to recognize in these kinds of semipermeable surfaces, but things can get far more elaborate.

For example, animals have immune systems that repair by fighting, neutralizing, and expelling internal intruders. Animals with neurons and brains also have mechanisms for avoiding and protecting themselves against predators as well as seeking and interacting with prey. As you know from personal experience, the human mind affords us the capacity to selectively interact with features of our environments, everything from dangerous food to dangerous people.

Still, all known mechanisms for selective interaction are evolved by natural selection. By definition, first selves have had no chance to evolve. Nonetheless, the first selves to emerge anywhere in the universe would have required selective interaction. How they would achieve it prior to evolution is a mystery we must solve, along with the mystery of how dynamics could ever transition from emergent regularization to emergent self-regeneration. As we'll see, Deacon offers a solution to this challenge.

19

OTHER EMERGENT REGULARIZATION DYNAMICS

OTHER KINDS OF EMERGENT REGULARIZATION

Before exploring Deacon's solution to the mystery of purpose, let's get beyond whirlpools to other examples of emergent regularizing dynamics. We'll do so for three reasons. First, as most origin-of-life researchers now acknowledge, self-organization (emergent regularization) is implicated in the origin of life, since selves are highly regularized.

Second, although emergent regularization is implicated at the origin, by itself, emergent regularization is never sufficient to explain the origins of selves. By visiting a few other well-understood examples of emergent regularization, often implicated in the origin of life, we'll see how emergent regularization fails where self-regeneration succeeds.

And third, Deacon's solution will rely on two additional emergent regulating dynamics that I must introduce: autocatalysis and crystal formation.

With this in mind, let's survey three other types of emergent regularization dynamics.

BÉNARD CELLS

Here's the recipe for Bénard cells, another oft-cited example of emergent regularization. Pour oil in a pan and heat it. As the oil heats, a honeycomb pattern develops on the oil surface, a regularization resulting from interactions throughout the dynamics. Heating the oil from below produces an energetic gradient, hotter oil on the bottom, and cooler oil on top.

The second law tendency toward irregularity equalizes this difference by conduction within the oil and thermal radiation into the air above and around the oil. But since the oil is heated continuously from below, the dynamics don't equalize completely. The bottom oil remains hotter than the top oil.

If the heat is turned high enough, the bottom oil gets much hotter than the top oil. When this temperature difference is large enough, the cooler top oil, being heavier, tends to sink, displacing the hotter bottom oil, resulting in rolling convection—cooler oil sinking and hotter oil rising.

Looking down into the pan, you would see the outlines of these rolls as irregular indented and raised areas across the oil surface. Gradually, these indented and raised areas regularize into a tight hexagonal pattern simply because a hexagonal shape maximizes the number of cells packed into the surface, thereby maximizing convection.

A whirlpool forms when water can't exit as fast as it enters, thus creating turbulent currents that then emergently regularize into a spiral flow. In like fashion, Bénard cells form when heat input exceeds the output possible by conduction and radiation alone. The larger heat gradient results in turbulent convection, which then regularizes into hexagonal convection cells.

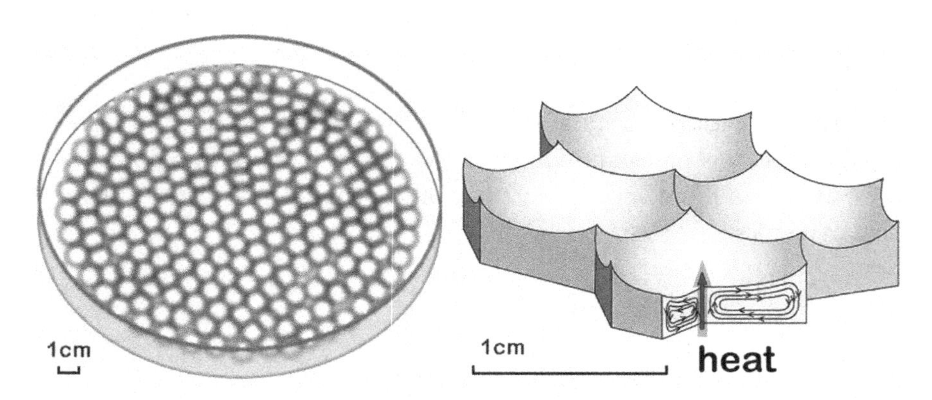

FIGURE 2 Bénard cells: Circular arrows indicate convection currents. Vertical arrow indicates heat dissipation currents.

FLASH IN THE PAN

With Bénard cells, nothing about the oil aims for regularity. It's just the second law equalizing temperature, as happened with our frozen pizza in the hot oven, but in this case, with a resulting emergent regularization—a path of least resistance for temperature equalization. In the case of hexagonal Bénard cells the process is convection—currents of oil—a more efficient means of dissipation than radiation and conduction alone.

Nothing about the oil aims for efficient equalization either. As with the whirlpool, it's just that once a system falls into efficient temperature equalization, it won't fall back toward a more inefficient equalization.

As with the whirlpool, the regularized currents aren't imposed from the outside, attracted from the inside, or somehow produced top down. Instead the regularization emerges from molecule-on-molecule work *throughout* the dynamics.

And as with whirlpools, at first there is no regularity present, and then increased regularity stabilizes into regularized hexagonal convection cells. If we increase the heat under the oil still further, the regular hexagonal cells collapse and the oil rolls chaotically or burns.

Like whirlpools, Bénard cells dissipate heat faster than it would otherwise dissipate and so are ephemeral. The hexagonal currents that emerge can do work to regularize things. For example, if you dropped small, heat-resistant rubber duckies onto the Bénard cell surface, they would fall into the hexagon's concave centers, rather than floating just anywhere on the surface.

But the work that Bénard cells do is not self-regenerative. There's no self-repair. When we reduce or increase the heat under them, the Bénard cell regularities simply become irregular and disappear. Nothing about the regularity counters the tendency toward irregularity. There's no self-protection. Bénard cells don't generate any insulation that resists the path-of-least-resistance heat exchange. In fact, they do the opposite; they make bottom-to-top heat exchange more efficient. And there's no self-reproduction. The work the regularized currents do does not somehow proliferate Bénard cell "offspring."

AUTOCATALYSIS

Catalysts are molecules that, due to their shapes and energetic properties, increase the likelihood of certain reactions occurring between other molecules (reactants), without the catalysts being affected in the process.

A catalyst's presence in a solution of reactant is a little like having a wheelbarrow for hauling gravel over a hill. Without the wheelbarrow, it takes a lot of work to move the gravel. With it, the gravel moves more efficiently and the wheelbarrow isn't altered in the process. Its presence therefore reduces work, haul after haul.

Catalysis is mere chemistry. There's no self aiming to get chemical reactions "over the hill." It is more apt therefore to think of catalysts as molecules that in the presence of reactants lower the energy required for a reaction to occur. They do so to varying degrees, and with varying catalytic products.

The presence of catalysts in solutions can increase the likelihood of reactants combining, for example, when catalyst X facilitates the bonding of reactants A and B into product D (A+B→D). Or the presence of a catalyst can increase the likelihood of reactants splitting, for example, when catalyst Y facilitates the splitting of reactant A into products B and D (A→B+D).

Catalysts often produce other catalysts in linear sequences. Catalyst X might turn reactants into catalyst Y. Thus, if you start with one catalyst X, you might end up with a slow-growing population of Ys all catalyzed by that one X.

And then there's *autocatalysis*, the sequence coming full circle, for example, one X that catalyzes reactants, thereby producing Ys, which catalyzes reactants, thereby producing more Xs. In such cases, there's a chain reaction, a chemical population explosion as reactants are converted to catalysts at an accelerating rate, not just Ys produced from one X, but a population explosion of Xs and Ys producing one another.

CREST AND CRASH

Catalysts' reusability makes autocatalysis something like the population explosion that would happen if people immortally reproduced forever, each offspring becoming another eternal offspring producer. The population

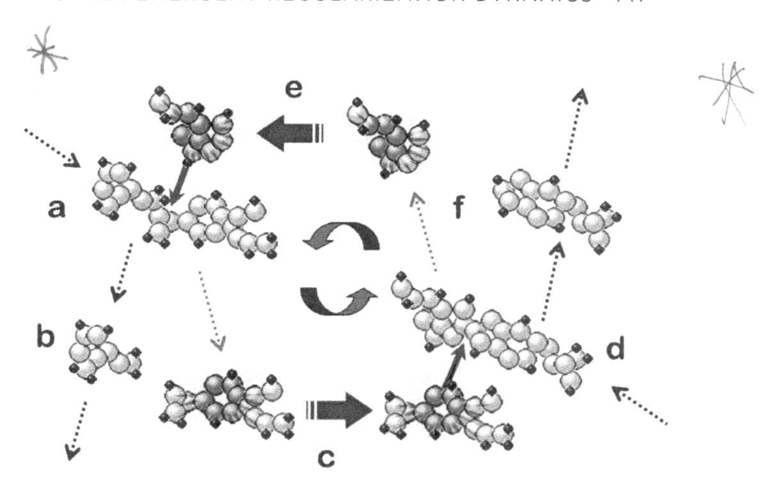

FIGURE 3 Simple illustration of autocatalysis: Molecule a is split into molecules b and c. Molecule c is a catalyst that splits molecule d into molecules e and f. Molecule e is a catalyst that splits molecule a into b and c. In this illustration, molecules a and d are reactants. Molecules c and e are catalysts forming the "autocatalytic set" and molecules b and f are autocatalytic by-products.

explosion would be expansive but also expensive, a rapidly growing population consuming reactants at an accelerating rate. In parallel, autocatalysis is expansive and will tend to deplete available reactants rapidly.

Unlike a human population explosion, autocatalysis isn't stuck in a limited environment. Catalysts would dissipate as they're being produced. Autocatalysis peters out when local reactants are depleted and catalysts are all dissipated. The *autocatalytic set*—in our example, catalysts X and Y—goes its separate ways, no more likely to restart autocatalysis elsewhere than any other individual loner catalyst.

To be clear, though scientists talk of autocatalytic sets, they do so just for analytical purposes. Nothing makes the molecules a real set. Throughout autocatalysis and after, it's all just individual molecules interacting. To think of the catalysts as forming a set is like thinking of a water current as a "water molecule set" when it's actually just individual molecules interacting.

Scientists recognize autocatalysis as another example of emergent regularization. But how? It doesn't produce a stable dynamic form like whirlpools or Bénard cells. Still, with the compounded production of Xs and Ys (often with byproducts, for example, if a catalyst splits a reactant), it results in an amplifying regularization. The more products are produced, the less likely it is that a reactant will escape being converted to a catalyst also.

Autocatalysis is like a population explosion, though strictly chemical, not a population of selves. And like a population explosion, it will tend to peter out, depleting the very reactants that the autocatalytic process depends upon. Stability is achieved briefly. While available reactants are being transformed, the population grows. But the population then collapses as local reactants become depleted and the concentration of local catalysts declines, catalysts dissipating, drifting away.

Autocatalysis does not have the three capabilities required for self-regeneration. It does not achieve self-repair. It poses no more resistance to its degeneration than a bathtub whirlpool poses to water draining out. It achieves no self-protection, generating anything that would keep catalysts from drifting away, dissipating in keeping with the second law. And though

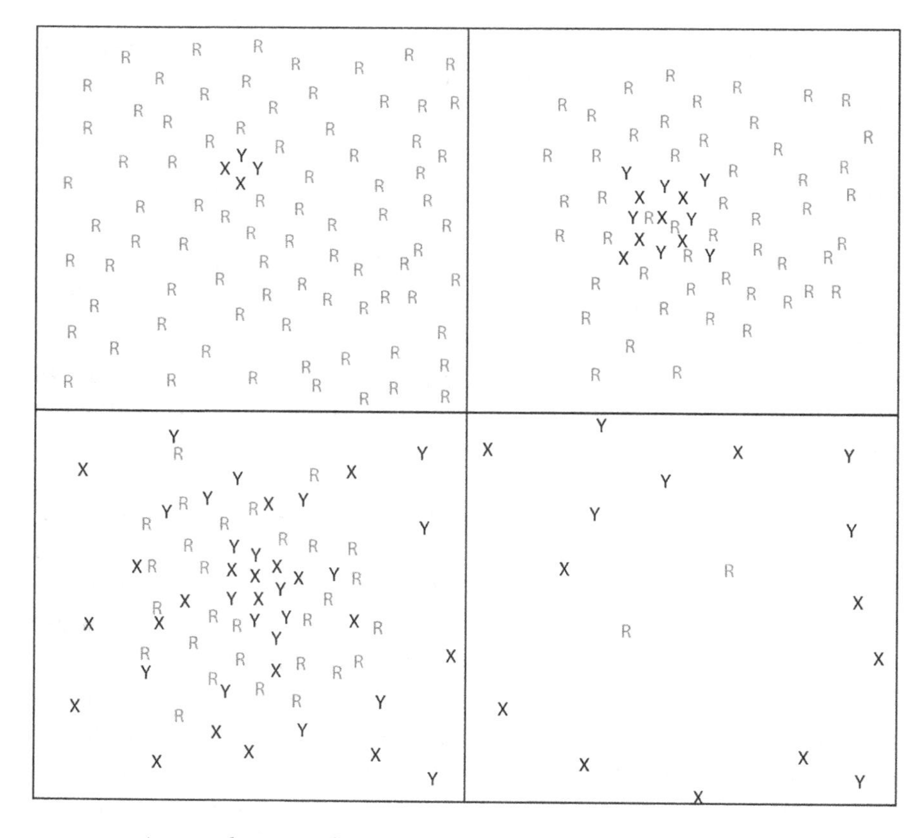

FIGURE 4 Autocatalysis. Catalysts X and Y (the autocatalytic set) transform reactants (Rs) into more Xs and Ys. Rs are depleted, and the Xs and Ys dissipate.

it generates a population of catalysts, it doesn't generate a population of autocatalytic dynamics, which are not really sets except in our analysis. Individual catalysts drift away individually, as likely to regenerate autocatalysis as any other individual catalyst might. Thus it does no self-reproduction: no reproduction of autocatalytic set offspring.

CRYSTALS: FROZEN REGULARITY

Our final example of emergent regularization is crystal formation, free-floating molecules binding into regularized structures. Crystal formation occurs across a vast range of materials, resulting in everything from polymers to rocks.

Many regularized structures in bodies assemble by some variation of crystal formation dynamics. These structures include cell membranes, microtubules, and molecular complexes such as the hemoglobin tetramers that ferry oxygen around in our blood.

Viral shells—spheres or tubes that spontaneously assemble from proteins—also form the way crystals do. They're called *capsids*. Though *capsid* only refers to viral shells, there's nothing about their formation that is exclusive to viruses. Due to their shapes and bonding characteristics, capsid molecules form shells by emergent regularization. As we discuss Deacon's model later, we'll refer to *capsid molecules* emergently regularizing, forming *capsids*—spheres or tubes—without implying their role in viruses.

Crystal formation in general is different from our other examples of emergent regularization for two main reasons. First, unlike whirlpools, Bénard cells, and autocatalysis, crystals can persist without an energetic gradient sustaining them. For example, a rock is a highly regularized crystal that can last for eons. Compared to rocks, capsid shells or tubes are relatively fragile. Some persist for a long time. Others are more fragile and can break when impacted.

The second distinctive feature of crystals explains the first. The molecules that form crystals fall toward regularization as energetic gradients equalize, whereas our other examples of emergent regularization only occur in the presence of energy gradients.

The molecules that form crystals are in an unstable state when they're free-floating, and are in a more stable state when regularized into crystals.

Crystal formation gives off energy whereas the other emergent regularizing dynamics require energy. This distinct feature of crystal formation is most intuitive when we think about how water becomes regularized into an ice crystal. We don't heat water to make ice crystal; we cool it.

When we drop water on ice, it freezes but it also warms the ice where it falls. That's the water giving off heat as it crystalizes, falling into a more energetically stable state. This is how snowflakes end up with their beautiful symmetry. Moisture freezing on one side of the flake warms it, making that side the least likely point for frozen moisture to accumulate next. Hence the snowflake crystal accrues ice on alternating sides.

We don't heat a solution to produce salt or sugar crystals; we cool it, eliminating the heat that is released as molecules relax into crystal conformations. In general, liquids become solids when cooled, not heated. Cooling liquids slows the molecules down enough that bonding into crystal lattice structures becomes possible. When they're heated instead, the crystal molecules are too active to bond into regularized forms.

Like all emergent regularization processes though, crystal formation is a *compounding* process. A salt crystal starts from a tiny seed with little surface area upon which free-floating molecules can bond. The more molecules attach to the crystal seed's surface, the more surface there is for other free-floating molecules to bond with. Free-floating molecules are thus increasingly unlikely to remain free floating.

With block crystals, the surface compounds indefinitely until the crystal reaches equilibrium given ambient temperature, with as many crystal molecules bonding as breaking free.

As capsid shells form from capsid molecules that bond edge to edge, their expanding edges make more bonding sites available. But when the sphere begins to close, bonding slows. With a fully closed sphere, all edge-to-edge bonds are completed and new bonding stops until the capsid breaks, for example, ruptured by bumping into other molecules.

As with our other examples of emergent regularization, with crystal formation there's no self-regeneration. There's no self-repair: increase the temperature and the crystal dissolves without putting up resistance to its disappearance. No self-reproduction: A crystal does not proliferate varied offspring. It simply grows. A piece of crystal can break off and grow elsewhere, but we wouldn't confuse that with self-reproduction, because there's no self-repair or self-protection transferred from parent crystal to offspring.

And though, as a static regularized structure, a crystal would make a good material for self-protection, it has no self to protect, since it has no capacity for self-repair or self-reproduction.

CRYSTALS AND LIFE

Though crystals are obviously not alive, their stable structures do make them important to origin-of-life research. In the same small book in which Schrödinger argued that life is a product of negative entropy, defying the tendency toward irregularity, he guessed that the only way life could pass on information from one generation to the next would be in the form of an "aperiodic crystal," a suggestion that eventually contributed to the discovery of DNA's structure, though DNA is not a crystal.

Aperiodic refers to a quality familiar from written language. When you write a word, it's static like a crystal, but the letters that make up the word can be connected to one another in any configuration. Any letter can be next to any other letter. There's no inherent bias in letter sequence. It doesn't take more work to put some letters together compared to others. This gives letter sequences the flexibility that enables us to use them to represent languages.

Schrödinger suggested that aperiodicity would also be a feature of the biological molecules that guide living activity, and DNA is indeed aperiodic. Its A, G, C, and T molecules bond to one another equally well in any sequence. Some researchers argue that DNA directly parallels language "read" by an organism's body. When we revisit interpretation in chapter 25 we'll explore the parallels and contrasts between language and DNA.

BRIEFLY COMPOUNDING CONSTRAINT

All emergent regularization processes regularize by compounding. The more people flow smoothly out of the foyer by some path, the more other people will be drawn into that path. The more water is swept into a whirlpool's spiral current, the stronger that current becomes and the more likely it is to sweep in other water molecules. The more oil molecules are drawn into Bénard cell hexagonal convection currents, the more likely other oil molecules will be pulled into the currents. The more reactants are converted into catalysts, the more likely that other reactants will be produced.

These descriptions suggest that emergent regularization results from compounding processes of production—the more of this, the more of that. Actually, all emergent regeneration processes result from processes of compounding elimination or constraint, the net-work that remains after possible paths for work to occur are netted out, constrained away by processes of elimination. What remains afterward are paths of least resistance.

The more that the paths through the foyer are constrained, the more the people will exit by the remaining paths. The more that turbulent water currents block one another, the more that the water will exit by the remaining unblocked paths. The more that heat can't equalize by conduction, the more that it will equalize through Bénard cell convection. The more reactants are converted to catalysts, the more that the remaining reactants are constrained into interaction with catalysts.

To get a better sense of the role that emergent regularization plays in autocatalysis, picture autocatalysis as like a rampant epidemic. The reactant molecules are like uninfected people, and infection is like being transformed into a catalyst. As the epidemic starts, people are free to go anywhere without risk of infection. As it spreads, fewer paths leave people uninfected, so more people become infected, which contributes to the increasing constraint on where people can go without infection. What drives the accelerating transformation is compounding emergent constraint—fewer and fewer paths available that escape infection.

In parallel, the midst of autocatalysis, paths by which reactants can remain untransformed are increasingly eliminated, and as a result, paths by which reactants are transformed into catalysts are the paths that remain.

THREE PROPOSED MISSING LINKS, ALL FALLING SHORT

Origin-of-life research has identified three primary distinguishing characteristics common to all organisms. Researchers tend to focus on one or another of these as the key to life's origin, since meeting multiple conditions would make life's emergence more unlikely. Thus, researchers tend to fall into one of three camps:

Metabolism first: Organisms grow, so life must have begun when populations of molecules grew, as with autocatalysis, a process that "consumes" or "digests" reactants, converting them into products.

Container first: Organisms are self-contained, so life must have begun with encapsulation, as with the self-assembly of capsids, or lipid (fat) or surfactant (soaplike molecule) bubbles forming tubes or spheres.

Information first: All known organisms maintain and pass on information molecules like DNA or RNA (called *template molecules*) so life must have begun when such molecules produced more of one another as with autocatalysis.

It's worth distinguishing these three criteria from the three that I have identified as essential for emergent self-regeneration:

Self-repair instead of metabolism: Metabolism or growth is not enough. Autocatalysis is metabolism, a growing population of chemicals, but it is not a self. Instead, our focus here on self-repair emphasizes what function growth serves, the self's maintenance and restoration of its regularities, not just the growth of a population of molecules.

Self-protection instead of containment: Chemicals within a container are not a self. Self-protection replaces the container-first requirement by focusing on the core functional aim of containment. Containment is the source of self-protection, the self's capacity to insulate or segregate itself apart from whatever environmental conditions would result in second law irregularity.

Self-reproduction instead of information: Information is an ambiguous term and is too often treated as an intrinsic characteristic of polymers that can replicate their aperiodic sequences the way DNA and RNA do. The fundamental aim served by self-reproduction is not the replication of polymers, but the ability to pass on a capacity to self-regenerate to offspring.

Metabolism, containment, and information are each seen by origin-of-life researchers as the possible breakthrough that led to the emergence of life, yet each is achievable through some variation on emergent regularization.

If life starts with metabolism or growth, then autocatalysis would explain life's origin. If life starts with containment, then crystal-like production of encapsulations would explain life's origin. And if life starts with the replication of inherently informational molecules, then catalysis or autocatalysis would explain life's origin, since, under rare circumstances, RNA molecules can act as catalysts in an autocatalytic set, that is, as one of the molecules X or Y in our illustration of autocatalysis.

Still, as emphasized earlier, emergent regularization processes do not achieve self-regeneration. They tend to peter out without putting up any self-regenerative resistance. No emergent regularization process achieves self-repair, self-protection, or self-reproduction.

In solving the mystery of purpose, our focus is on the self for which metabolism, containment, and information can be about something for the self's aims. Any model that merely focuses on production of a self's components will not prove sufficient to solve the mystery of purpose.

Selfhood does not begin with autocatalysis, the production of containers, or the replication of the kind of aperiodic molecules that in evolved selves bear information. Instead, selves emerge as a phase transition from emergent regularization to emergent self-regeneration, self-directed work aimed by the self for the self.

Since no single emergent regularization process is capable of self-regeneration, our next step is to explore possible *synergistic couplings* between emergent regularization processes. Perhaps in combination, two or more emergent regularization processes might constrain or prevent each other's tendency to peter out.

20

COUPLED REGULARIZATION PROCESSES

RECIPROCAL MEANS AND ENDS

In 1790, philosopher Immanuel Kant hinted at just such a synergistic coupling as the distinguishing feature of organisms. In his *Critique of Judgment*, Kant distinguished between a machine's *motive power* and life's *formative power* (something akin to self-regeneration). Kant argued that with formative power "Every part is thought as owing its presence to the agency of all the remaining parts, and also as existing for the sake of the others and of the whole, that is, as an instrument, or organ. . . . The part must be an organ producing the other parts—each, consequently, reciprocally producing the others."[1] He goes on to say that "The definition of an organic body is that it is a body, every part of which is there for the sake of the other (reciprocally as end, and at the same time, means)."[2] Kant, the ever-skeptical inquirer, acknowledged that his observation of the reciprocal means-and-ends relationship might simply be an eye-of-the-beholder impression. Thus, he left uncertain whether organic bodies truly differ from nonliving systems, or only in the impression that we have of them.

Kant wrote his account long before the discovery of self-organization (emergent regularization). Updating his insight, we can imagine that in a prebiotic universe, Kant's reciprocal means-and-ends relationship might have resulted from the chance *synergetic coupling* between two emergent regularization processes, for example, between two autocatalytic processes or between an autocatalytic process and a crystal-forming process.

I'll call any interaction between emergent regularization processes a *coupling*, but a coupling by itself is not enough to achieve Kant's reciprocal means-to-ends relationship. What matters for our exploration is whether the coupling is *synergistic*, in other words, whether coupled emergent regularization processes constrain one another's tendencies to peter out, thus achieving synergy (literally, *working together*), Kant's reciprocal means-to-ends relationship.

The synergy necessary for the emergence of selves would not be something added, or some top-down whole somehow becoming more than the sum of its parts. Rather, it would be a way that emergent regularization processes would constrain one another, thus reducing the likelihood of both petering out as the other would independently of one another.

You will recall that Deacon turns our intuition about synergy on its head, arguing that the whole is *less* than the sum of its parts or, more accurately, less than the sum of all possible dynamic paths. That is the definition of synergy that I'll employ here in seeking a synergistic coupling between regularization processes.

The synergistic coupling that would achieve self-regeneration would be the result of a process of elimination, a reduction in dynamic paths. Thus, emergent self-regeneration would be a higher-order emergent constraint resulting from the synergistic coupling between two lower-order emergent regularization processes, each acting as means to constrain the other's tendency to peter out. This higher-order emergent constraint would be a further constraint upon the emergent regularization processes' underlying constraints.

We will visit three candidates for a synergistic coupling, the first a coupling between autocatalytic processes, the second a coupling of autocatalysis and crystal formation, and the third autogenesis, Deacon's solution to the mystery of how selves and aims can emerge from aimless chemistry.

HYPERCYCLES

Chemist Manfred Eigen proposed a model worth exploring as a possible candidate for synergistic coupling. He called it a *hypercycle*, a coupling of two autocatalytic processes, each contributing catalysts to the other.[3]

To illustrate, imagine two autocatalytic sets, one consisting of catalysts A, B, and C, and the other consisting of catalysts C, D, and E. Each of these two sets produces more of the catalysts in the set. For example, with the first set, catalyst A transforms reactants into catalyst B, which transforms reactants into catalyst C, which transforms reactants into catalyst A. Reactants are depleted as a population explosion of As, Bs, and Cs grows and dissipates. Nearby in parallel, the second set is depleting reactants while producing a population explosion of Cs, Ds, and Es.

With the two autocatalytic processes interacting, and therefore coupled with each other, the Cs generated by both cycles would be plentifully available, which could make the hypercycle more productive than a single autocatalytic process.

If the hypercycle persisted, perhaps a third autocatalytic process would become involved, for example, with a set consisting of catalysts E, F, and G, thus providing a second source of Es and making the whole hypercycle still more productive. Eventually the hypercycle would accumulate enough coupled autocatalytic sets that it becomes sustainable.

If this hypercycle model explained the origin of selves, it would vindicate thresholdism by demonstrating that selves could emerge simply through increased interaction between more catalyst types in an autocatalytic process.

But hypercycles are unsustainable due to what's called *error catastrophe*, instability that increases as autocatalytic processes become more intricate, involving more steps, and molecular types.

ERROR CATASTROPHE

To get a sense of error catastrophe, consider the complications that arise in very intricate manufacturing or building processes. If the steps aren't coordinated and sequenced well, they fail. In manufacturing and building processes, error catastrophe can be prevented through deliberate managerial constraint—human selves aiming to sequence, schedule, and orchestrate the intricate steps in the production process. With hypercycles, there's nothing to constrain the sequenced steps so the growth and sustainability we can imagine when there are more steps become illusory.

In our illustration, the hypercycle might rapidly overproduce a concentration of Cs that then interfere with other reactions, in loose parallel to the way that overproduction of some element in a manufacturing process can clog the factory.

A hypercycle is not enclosed in anything. It's not like a factory. The autocatalytic sets are just a local concentration of independent molecules in chance interaction. Indeed, the two sets ABC and CDE might just as well be called set ABCDE.

The more catalysts and reactants are involved, the greater the likelihood that some will be over- and undersupplied in proportion to one another, some dissipating before they catalyze the other members of the set. This loosely parallels an intricate food manufacturing process in which one perishable ingredient is overproduced and perishes before it can be used, or another ingredient becomes unavailable, stalling production.

Besides, a hypercycle is not really a coupling. It's just a single and more intricate emergent regularizing process. Given that larger autocatalytic sets increase the possibility of unconstrained error catastrophe, a hypercycle is not a synergistic coupling. The potential for error catastrophe argues not for but against thresholdism as the origin of self-regeneration.

Hypercycles are not self-regenerative. There's no more capacity for self-repair, self-protection, or self-reproduction than there is with autocatalysis. Even if error catastrophe didn't occur and the many steps in the process were to run in an efficient sequence, a hypercycle would just be a faster, larger, uncontained molecular population explosion. Reactants would be depleted more rapidly as they were transformed into catalytic products. The concentration of catalytic products would dissipate as they do with any autocatalytic process. Lacking the capacity for self-regeneration, a hypercycle does nothing to constrain the second law tendency to peter out.

Hypercycles are not synergistic couplings. Indeed, they are hardly even couplings in that combining autocatalytic processes merges them into one. Let's next explore a coupling of two different kinds of emergent regularizing processes, autocatalysis and the crystal-like formation of containers.

CONTAINER-GENERATING AUTOCATALYSIS

Imagine autocatalysis occurring within a container. It would look a little like a very simple cell—a chemical population explosion occurring within something like a cell membrane. To enhance the coupling, imagine that one of the autocatalytic by-products was a capsid molecule, the very kind of molecule that regularizes, yielding a container.

In my earlier illustration of autocatalysis, catalysts X and Y transform reactants into more of each other. Some catalysts do so by facilitating the bonding of reactants to each other. Others do so by splitting reactants apart. So, for example, catalyst X might split a reactant molecule in two, one of the resulting molecules being catalyst Y, the other being a capsid molecule or perhaps a lipid or a surfactant, molecules that, in a process like crystal formation, regularize, yielding containers.

This model is something like autocatalysis in a bottle, but more than that, like autocatalysis that generates the molecules necessary for the production of the bottle.

By origin-of-life research standards, this model is intriguing. It appears to satisfy the metabolism-first standard since the autocatalysis is the source of growth, transforming or "consuming" reactants, turning them into its own molecular types. Through container production it satisfies the container-first standard. If one of the catalysts in the autocatalytic set were an RNA molecule, it might even satisfy the information-first standard in that it would produce more RNA, though we have yet to see how RNA or any molecule becomes information. For the moment, we'll set RNA aside.

Chemist Luigi Luisi explores contained autocatalysis both theoretically and experimentally and proposes two main models. For both models he imagines surfactants (like soap) or lipids (fats), which produce bubbles regularizing to form containers—membranous vesicles—in a process much like crystal formation. Luisi's first model relies on a ready external supply of surfactant or lipid molecules from the environment since, as Luisi argues, "It is likely that during the chemical evolution leading to the first catalytic and replicating molecules, the ancestors of today's proteins and nucleic acids, membranous vesicles were available in the prebiotic environment, and ready to provide a home for the first forms of cellular life."[4]

With the second model, which will be our focus now, autocatalysis contributes to the production of surfactant molecules that regularize, forming containers in a process somewhat like crystal formation. As described by philosopher Evan Thompson, who collaborates with Luisi, this container-generating autocatalysis is "a bounded structure that hosts in its aqueous interior a chemical reaction that leads to the production of the surfactant, which in turn aggregates to form a boundary for the reaction."[5]

Luisi and Thompson subscribe to a definition of a living unit as "autopoietic," or self-producing. As Luisi defines it, "An autopoietic unit is a system that is capable of sustaining itself due to an inner network of reactions that regenerate the system's components." Thompson claims that Luisi's models represent "A minimal autopoietic system."[6] Before considering autopoietic theory, let's explore these autopoietic units in their own right.

EXPECTING MORE LUCK THAN IS LIKELY

In origin-of-life research, metabolism suggests growth, the conversion of externally available resources into growing material bodies. This provides expansion, but also a way to replenish lost material.

Would the autopoietic unit grow? Presumably it would through autocatalysis, which, as we've seen, is something like an accelerating chemical chain reaction. But this chain reaction grows only through interaction with reactants that the autocatalysis transforms into products.

This presents a problem since the container that holds the autocatalytic set together is likely to prevent interaction with reactants outside it. This is the *selective interaction* problem described earlier: self-protection requires containment, which prevents interaction, but self-repair and reproduction require interaction with energy and resources.

To solve this problem, Luisi proposes that the autopoietic unit would have a selectively semipermeable membrane. As Luisi describes it, "Here the term semi-permeable means that certain substances (nutrients and other chemicals) are able to penetrate in the interior, whereas most other chemicals cannot. This is a kind of chemical selection and chemical recognition—a notion that will be used . . . in connection with the term 'cognition.'"[7]

Cognition is to Luisi what a self's aims are to Deacon. "Cognition," as Luisi uses it, is expanded well beyond its conventional mental connotation, encompassing any functional capacity for selective discernment by an organism of its environment.

To Luisi, the origin of cognition and therefore of life is found in a selectively semipermeable membrane, a container that permits useful, but prevents harmful, interactions. I touched on this earlier as addressing the problem of selective interaction, and noted that, indeed, all known lifeforms solve the problem by means of selectively permeable membranes. I also noted that it is highly unlikely that even modestly functional selective permeability would arise by chance in prebiotic chemistry.

Autopoietic units, as Luisi imagines them, would be autocatalysis occurring inside selectively permeable membranes all emerging by chance, but nonetheless functioning within very tight tolerances: Just the right autocatalysis generating just the right container-molecule by-products, which aggregate into just the right selectively permeable containers that would permit just the right selective interaction. It is very unlikely that all of this would come together by accident alone in chemistry.

Still, if chance were with it, the whole autopoietic unit might grow. This growth is where Luisi and Thompson see some potential for reproduction (although reproduction is not a requirement in their definition of autopoiesis, which emphasizes self-production, not self-reproduction).

In their model, the autopoietic unit's membrane would expand and possibly split in an extremely loose analogy to cell division. Membrane expansion would require the production of a large quantity of surfactant molecules, which is the reason that Luisi suggests that the autopoietic unit might require a ready external supply of surfactant molecules sufficient to keep pace with the growing autocatalysis.

CLOSED, BUT NO SOLUTION

Assuming that somehow all of the feasibility problems could be overcome, and autopoietic units did emerge, would they be capable of self-regeneration?

They wouldn't self-repair. Luisi counts on the autopoietic unit's ability to "regenerate the system's components." But supplying component molecules is not the same as self-repair to the whole unit. In fact, even autocatalysis alone generates the system's components. It's just not a system really, not a unity, but rather individual molecules interacting. The autopoietic unit does no work to maintain its unity. If broken open, its contents simply dissipate.

Is there self-protection? There is, but given that there is no self-repair, it is ephemeral. If the enclosure were broken, there's some chance that the disparate lipid molecules would reform but not through self-repair work done by the autopoietic unit's own work. Self-protection isn't just the existence but the regeneration of protection.

The surfactant container holds the autocatalytic set together, but only if it isn't broken. If broken, the dispersed catalysts lose the very proximity that allowed them to autocatalyze, and therefore the capacity to produce the byproduct lipids required for recontainment.

Does the autopoietic unit self-reproduce? In theory it could by growing, expanding, and splitting, perhaps very loosely analogous to mitosis. But when present-day cells divide, they first produce duplicate copies of their self-regenerative mechanisms and divide them in ways that ensure that daughter cells each get a copy. Self-reproduction only works when offspring inherit the self-regeneration capabilities of parents, not when mere haphazard splitting occurs.

UNITY IN THEORY ONLY

According to philosopher Francesco Varela, codeveloper of the autopoietic theory, "For a system to be autopoietic, (i) the system must have a semipermeable boundary; (ii) the boundary must be produced by a network of reactions that takes place within the boundary; and (iii) the network of reactions must include reactions that regenerate the components of the system."[8]

Within autopoietic theory, life's defining feature is unity, not selfhood and aims. By most autopoietic theory accounts, an autopoietic system is a materially coherent unity that facilitates the production of molecules that are components of the material unity. But autopoiesis theory prioritizes unity without explaining how it is achieved. What it ignores is that given the second law tendency toward irregularity, unity has to be maintained actively through self-regeneration, the self's constrained work to keep itself together.

Life-forms as we know them are self-contained from embryo to death. Uninterrupted containment like this makes the unity challenge seem easy to solve. Just bottle up the protoplasm and it becomes a unity. But containers break down due to the second law, so the molecules that aggregate to produce the containment would have to be resupplied. Still, whether the protoplasm (in this case, autocatalysis) replenishes molecule supply or they just happen to be available externally as Luisi's first model suggests, a ready supply of molecules is not enough to claim a capacity for generating its own unity.

With autopoietic units, the ephemeral unity achieved through containment is of low value at high cost: low value because it is easily lost—the unit has no capacity to repair a ruptured containment—and high cost due to the extraordinary luck necessary to achieve the highly unlikely selective permeability that Luisi counts as the source of its distinctive selective interaction. Given the requirements for selective interaction, the containment feature is a bug, a design flaw that makes it unlikely to emerge and unlikely to persist.

Thus the autopoietic unit is a coupling of emergent regularization dynamics—autocatalysis and a lipid membrane dynamic that is similar to crystal formation, but it is not a synergistic coupling, a coupling that constrains paths such that self-regeneration emerges.

Still, autocatalysis that produces by-product container molecules is a potentially promising coupling. It couples distinct kinds of emergent regularizing dynamics: autocatalysis, which requires energy, and crystal formation, which gives off energy. If only there were ways to overcome the bugs.[9]

And it turns out that there are.

V

DEACON'S SOLUTION

21

AUTOGENS

Self-generators

BUGS BECOME FEATURES

As an origin-of-life model, the autopoietic unit has two main bugs: first, to be productive while contained, autocatalysis requires a highly improbable selective permeable membrane; and second, nothing about its dynamics repair containment if the container is broken. Thus, it does no work to maintain its own unity.

But what if, unlike all known selves, the first selves didn't have to be consistently contained? If so, we could overcome these bugs—indeed, turn them into features. Picture, then, Deacon's model for the first selves, which he calls *autogens* (self-generators).

Like the autopoietic unit, the autogen is a coupling of autocatalysis and container formation. Its autocatalysis produces, as a by-product, capsid molecules that, through emergent regularization, assemble into capsids, shells like those that we find as viral capsules today.

Unlike the autopoietic unit, the autogen is closed sometimes within the capsid shell, and open at other times. When contained, no reactants enter or exit and the autocatalytic set is dormant. When the container is broken open in the presence of reactants, autocatalysis resumes, producing more catalysts and capsid molecule by-products. These by-product molecules regularize, forming capsid shells amid the autocatalytic process. Some of these shells would encapsulate a sampling of the locally produced members of the autocatalytic set.

While the autopoietic unit depends on autocatalysis going on within an always-closed container, the autogen does not. Inside the closed autogen the catalysts are dormant. With the autogen, autocatalysis occurs only when the container is broken, just where and when second law tendencies are likely to dissipate the molecules that recontainment depends upon and therefore just where the regeneration of catalysts and capsid molecules is most beneficial.

In the autogen model, the lack of a selective permeable container is not a flaw but a feature crucial for self-protection, a way in which the dynamics can lock down a dormant sampling of the autocatalytic set that can persist even in nonsupportive environments lacking reactants.

Since, when open and active, autocatalysis might produce a lot of capsid molecules, and therefore capsids, it is able to self-reproduce proliferating varied "offspring," in that the multiple containers are likely and each could encapsulate a varied sample from the catalytic set and possibly other molecules sampled from the environment.

With the autogen, the challenge of achieving selective interaction is solved not with highly improbable selective permeability arising by chance in an abiotic environment, but through its alternation between open and closed phases. The open phase achieves self-repair and reproduction. The closed phase achieves self-protection.

In the closed phase, the encapsulation and its contained catalysts would be a bit like dormant seeds. I'll call them seeds here, though keeping in mind that this is a metaphoric use of the term. The seeds would contain a varied yet potentially representative sampling of the autocatalytic set. If

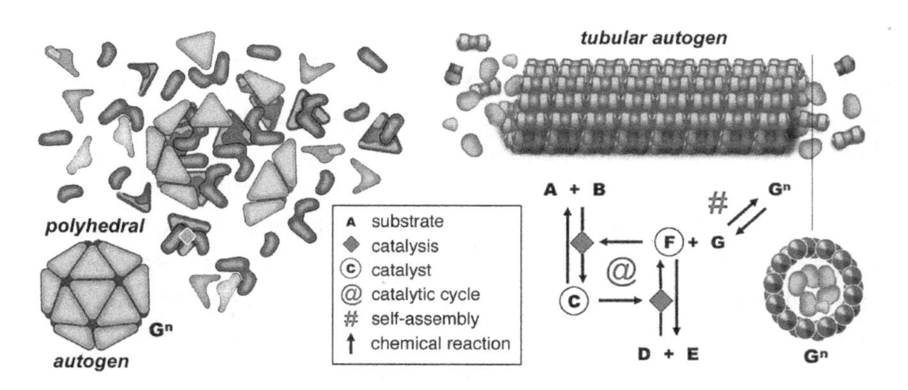

FIGURE 5 Spherical (left) and tubular (right) autogen models with chemical interaction diagrams.

broken later in the presence of more reactants, autocatalysis would resume producing yet more seeds.

By alternating between an open and a closed phase, the autogen builds up and locks down its regularities, repairing and reproducing them for a time and then protecting them in seeds. Autogens would be evolvable. They would have heredity and variation and would thus be subject to natural selection.

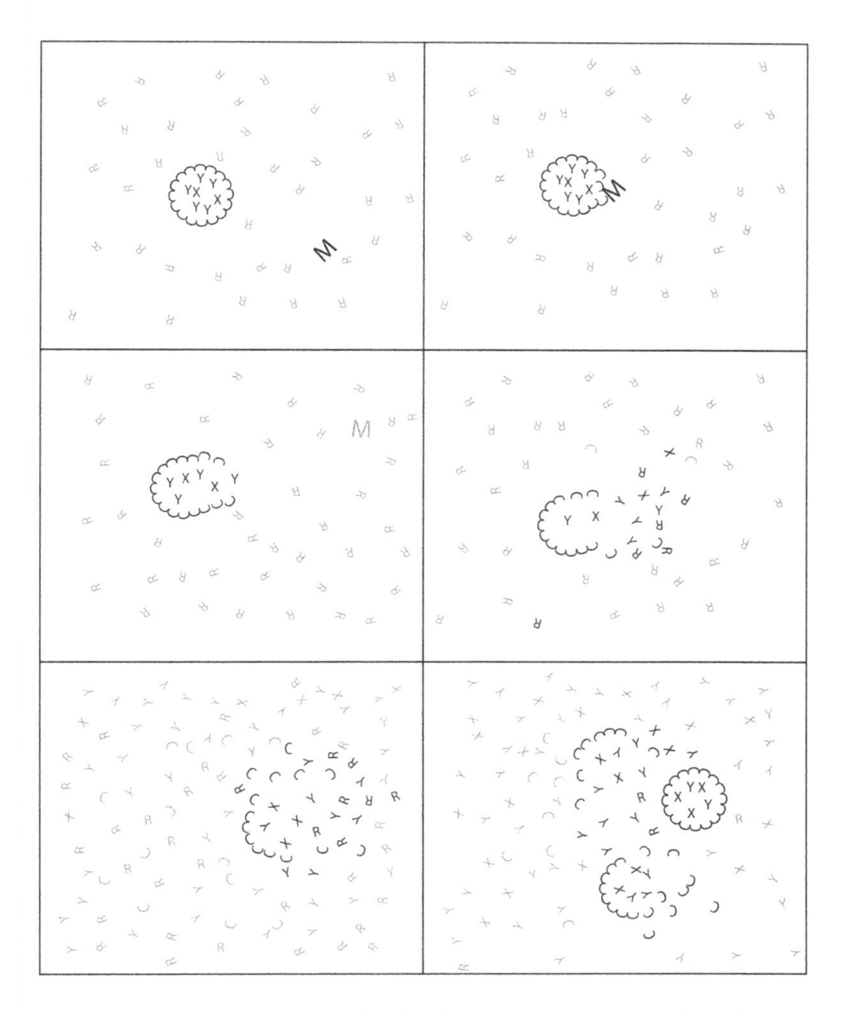

FIGURE 6 Minimal autogen. When broken by chance interaction (with large molecule M), autocatalytic set X and Y converts available reactants into more Xs, Ys, and Cs (capsid molecules), which regularize into new "seeds," encapsulations containing varied samplings of Xs and Ys.

To illustrate, imagine autocatalytic set X and Y converting reactants into more Xs and Ys plus capsid molecule by-products C. Right where autocatalysis is generating lots of Xs and Ys. Capsid molecules would self-assemble into capsids containers. Some of these containers would end up encapsulating varied samplings of catalysts X and Y. Local reactant would be depleted by autocatalysis but not before seeds were produced.

Seeds might break open again elsewhere, or in the same area once reactants had been replenished by diffusion from elsewhere. They would break by happenstance, chance interactions with things outside the encapsulations. When they opened, if reactants are present, autocatalysis would resume, regenerating and replenishing catalysts and capsids, thus producing more seeds.

Deacon argues that an autogen is a synergistic coupling that achieves true self-regeneration, thereby providing a testable proof of concept that it is possible to solve the mystery of purpose relying solely on conventional physical science. What follows will be his argument for how and why this is the case.

NOT THE MATERIAL OBJECT BUT THE CONSTRAINED DYNAMIC TENDENCIES

The autogen cycles between two phases, open and closed, due to the synergistic coupling between the two emergent regularization dynamics—autocatalysis and capsid formation. In the open phase, autocatalysis regenerates seeds. Since autocatalysis is far more likely to restart from a cluster of catalysts than from an individual catalyst, containment constrains the likelihood that autocatalysis will end with catalysts dissipating, never to autocatalyze again.

Given materialist intuitions, it might seem right to identify only the encapsulated catalysts as the autogen. After all, the seed is a material object that most resembles a body. Or one might assume that an autogen can't be a self since it's not continuously contained. When open, it's nothing but a constellation of independent catalysts.

When seeds break open, there is just a loose constellation of independent molecules. This may seem like nothing more than autocatalysis but it's actually autocatalysis that, due to the constraining effects of capsule formation, tends toward recontainment.

The autogen thus self-regenerates by means of its cycle: open self-repair and self-reproduction tending toward closed self-protection, and closed self-protection tending toward open self-repair and self-reproduction. An autogen is neither the closed nor the open phase but the tendency to close when opened and open when closed.

In very loose parallel, a sunflower isn't the seed phase or the plant phase but the complementary tendency to alternate between the phases. The autogen is even a little like the chicken and egg. Regardless of which comes first, we identify the pair of alternating phases as a self within a lineage of selves.

The autogen is the tendency when closed a self to open, and when open to close. In other words the autogen *self* is the tendency to self-regenerate by means of a constrained cyclic tendency.

AUTOGENS SELF-REGENERATE

In the autogen, we meet all three criteria for self-regeneration, albeit in a most primitive form. There's self-repair in the way a broken enclosure recloses. There's self-protection in the static seeds. And there's self-reproduction in the way that, when open, multiple seeds are likely to form, each with its varied and therefore evolvable sampling of encapsulated catalysts, and each with it own chance of breaking open in the presence of reactants.

sums up

By alternating between an open and a closed phase, the autogen achieves selective interaction. It does so in a way that our two other coupled models—hypercycles and autopoietic units—do not. Autogens are not always open like hypercycles, or always closed like the autopoietic unit. Instead, autogens are sometimes open and interacting with local energy and resources, channeling them into self-repair and self-reproduction, and sometimes closed and therefore self-protecting.

In a papers published in 2006 Deacon originally called the autogen an "autocell" but renamed it because the term *autocell* implied cell-like containment. An autogen is, in contrast, loosely analogous to a nonparasitic virus.

Viruses are not cell-like in that they are enclosed in an impermeable capsid shell. Because of this they are closed and inert when outside of a host cell. When viruses enter host cells, their contained DNA or RNA molecules become freed. During this phase, the viral DNA or RNA molecules are multiply copied by the host cell and are also used to synthesize more viral capsid proteins. As a result, new capsids form, enclosing the newly synthesized viral DNA and RNA.

Whereas viruses depend upon host organisms for regeneration of the inert viral phase, in the autogen, the contained molecules are catalysts that transform reactants into more catalyst and capsid molecules that spontaneously reform into new inert autogen "seeds."

The autogen does not open when it is advantageous to do so; rather, it opens when bumped and broken, not necessarily when reactants are present. Breaking in the absence of reactants, autogens would fail to repair and reproduce. The catalysts would merely dissipate, the end of the line for that lineage of autogens. Shortly, we'll show how a *selective autogen* could evolve, an autogen that is more likely to open when reactants are present.

Because we will explore evolved variations on the autogen, we'll call the most basic model the *minimal autogen*. It merely alternates between interacting and not interacting. Still, this capacity sets up the conditions for evolvability. More fundamentally, it makes self-regeneration possible.

THE MINIMAL AUTOGEN

In the minimal autogen we find metabolism and enclosure, two of the three features that different schools of origin-of-life researchers argue must be present at life's beginning. And what about the third feature, information-bearing molecules?

Deacon agrees with the information-first approach to the origins of life only insofar as he agrees that life and a capacity for interpretation emerged together. Even before evolution, first selves would have had a minimal capacity to interpret or be about their circumstances.

The capacity for interpretation in the minimal autogen is indeed minimal, and we shouldn't expect more from a first self. The information does not take the form of a material object, a template molecule like DNA or RNA, but rather exists in the autogen's tendency toward self-similarity from moment to moment and generation to generation.

The most easily intuited interpretive relationship in the minimal autogen is in what it does when the capsid breaks. In response, the autogen resumes autocatalysis, replenishing catalysts and capsid molecules that regularize, regenerating autogens.

Whatever external force breaks the capsid doesn't *cause* this response, any more than a red light causes your foot to hit the brake or sugar pulls the bacterium toward it. The autogen's interpretive response is a function of its tendency to close when opened.

The autogen emerges as a chance physical configuration, the synergistic coupling of its two regularizing dynamics, autocatalysis and capsid formation. Still, it's a synergistic coupling that happens to maintain the synergistic coupling. That it happens to do this provides the basis for describing its behavior as both means and ends. Its means are the synergistic coupling; its ends are the maintenance of that synergistic coupling.

The autogen's synergistically coupled dynamics *constrain* work into the maintenance of its synergistically coupled dynamics. Deacon argues that the synergistic coupling is both the self and the aim, which may seem highly counterintuitive given our materialist habits of thought, but may become more intuitive as we explore further.

INFORMATION FOR THE SELF ABOUT ITS ENVIRONMENT

Intuitively, we would say that what is passed on through autogen generations is information, not carried in template molecules, but in its modest capacity to interpret its environment, its modest fittedness to or about its circumstances.

The minimal autogen's tendencies are about its circumstances only in the broadest possible sense. Its ability to protect its regularities when closed is vaguely about the intermittent availability of reactants in its environment. Were reactants always in infinite supply, there would be no functional advantage to closing and no benefit therefore to producing capsid molecule by-products. Autocatalysis could run continuously, though without any self-regenerative features since, as we have seen, autocatalysis is merely an emergent regularizing dynamic.

Conversely, if reactants were never available, or if nothing could ever break capsids, there would be no advantage to protecting a cluster of catalysts in a capsid that when broken could resume autocatalysis, and such an entity would not be a candidate for a protolife form. The autogen is viable in, and therefore fitted to and representative of, an environment that includes intermittently available reactants and the chance of breakage.

The autogen's cyclic tendencies—basically, a protolife cycle—represent its circumstances and re-presents its dynamics over time and generations. Whereas emergent regularization dynamics such as whirlpools have no conditions that they work to maintain, the autogen's self-regeneration works to maintain its ability to self-regenerate. Within a range of possible environmental circumstances, the autogen's tendencies reproduce its own tendencies. It maintains self-similarity over time, each moment of its existence representative of prior moments. It *remembers* itself.

The autogen remembers its dynamics in two senses: first, through repair and protection, which protects and restores its constraints when they're disturbed, and second, through reproduction, in other words, re-membering or repopulating its environment with evolvable members of its lineage.

SELF-CLEANING

In real-world chemistry, minimal autogens would not exist in a tidy environment where reactants are either present or absent, but rather in an environment teeming with various molecules, some of which could impede the encapsulation or autocatalytic processes. Some of these contaminating molecules would bond or react with catalysts, taking them out of action.

Some molecules might simply get in the way, blocking access to reactants, thereby slowing autocatalysis. Capsids would likely enclose a sampling of these contaminating molecules. Thus, an autogen is threatened by an accumulation of deleterious molecules that, through error catastrophe, would reduce its capacity for self-regeneration.

But even the minimal autogen has some capacity to purge deleterious molecules. Most interloper molecules wouldn't contribute to autocatalysis, and that proves to be good news for autogen repair. When a seed breaks in the presence of reactants, its enclosed catalysts produce copies of one another, but the contaminating molecules do not. As a result, subsequently produced seeds are far more likely to contain catalysts than counterproductive molecules. For example, if a stray molecule J was enclosed with catalysts X and Y, when the resulting seed next breaks and reseals in the presence of reactants, catalysts X and Y would reproduce, whereas contaminant J would not. Counterproductive stray molecules would come and go, while catalysts would stay well represented, given their replenishment by autocatalysis. In this sense, autogens are self-cleaning or self-purging each time they open, another self-repair feature.

In the minimal autogen's capacity for self-purging, we see perhaps life's earliest example of another important idea that Claude Shannon contributed

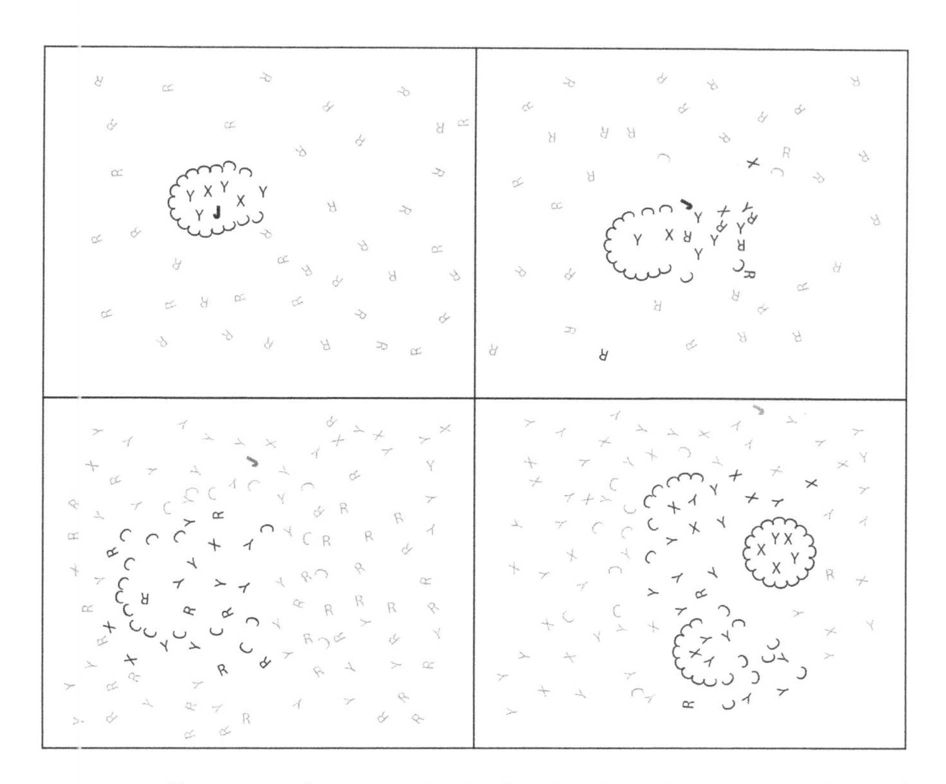

FIGURE 7 Self-purging. Deleterious molecule J doesn't replicate through autocatalysis and therefore is unlikely to be inherited in next-generation autogens.

to information theory. Shannon argued that noise can be purged through redundancy, the repetition of a communication. For example, when a cell phone signal is breaking up, one can determine what is signal and what is noise by repetition, because the noise is variable while the signal remains the same. If one repeats, "I'll be there" over a noisy channel, it might be received as some variation on "I'll . . . there," " . . . be there," "I'll be. . . . " Through repetition or redundancy, the message can still be interpreted.

In parallel, with the autogen the noise of counterproductive molecules varies from generation to generation. They come and go, but autocatalytic molecules persist redundantly, a steady signal amid random, transient noise.

EVOLVABLE REPRODUCTION AT THE EDGE OF CHAOS

I noted earlier that catalysts aren't all-or-nothing causes of reactions. Rather, they increase the probability of certain reactions to varying degrees. As such, there's potential for substitutions that might increase autogen efficiency.

Since autogens are sometimes open, their seeds might come to contain not just different quantities of catalysts X and Y, but perhaps substitute catalysts that might be present in the local environment. A lineage of autogens might end up with a substitute for catalyst Y that makes autocatalysis more efficient.

For example, a lineage with seeds that carry forward samplings of catalysts X and Y might at some point include a substitute for Y—say, catalyst Z—which might be more efficient than catalyst Y in transforming reactants into catalyst X. Substitute catalyst Z would not be purged when capsids break. The evolved substitution of catalyst Z for catalyst Y might even result in a different capsid molecule by-product, and thus in an improved variation on the original capsid containment (a possibility explored next in Deacon's selective autogen model). Catalyst substitution contributes to evolvable variation in autogen lineages.

Evolutionary theorists recognize that one can have too much or too little variation from generation to generation. Too little variation and the possibility for evolved adaptation is compromised; too much and the lineage becomes unstable. Dynamic systems theorists often describe life as lived "on the edge of chaos,"[1] meaning that with too little variation, life can't evolve, whereas with too much, it can't survive. The autogen's reproductive balance between repair and protection enables it to explore variation without drowning in it.

22

EVOLVED AUTOGENS

THE SELECTIVE AUTOGEN

Deacon's minimal autogen has the capacity to interact intermittently with its environment, but it isn't selective about when it interacts. It's open when broken by chance, regardless of whether the environment contains the reactants necessary for self-repair and self-reproduction.

Evolving selective interaction would be highly adaptive, a capacity to open when reactants are present and to otherwise stay closed. Deacon imagines how such a capacity might evolve. We will call this evolved variation a *selective autogen.*

Deacon imagines a minimal autogen lineage that through catalyst substitution produced a variation on the original capsid that opens more readily in the presence of reactants. For example, he imagines a capsid surface that tends to bond to reactants, making the surface rougher and therefore more likely to catch on other molecules, thereby increasing the chance of breaking. If the reactants bonded across several capsid molecules, they might break free when the capsid breaks.

An evolved autogen that was more likely to break open in the presence of reactants would have internalized a capacity for true selective interaction. It wouldn't just break open by chance but to a modest extent select when to open, not by choice but by an evolved mechanism, an ability to distinguish functionally between advantageous responses to two different kinds of environments—environments with and without reactants.

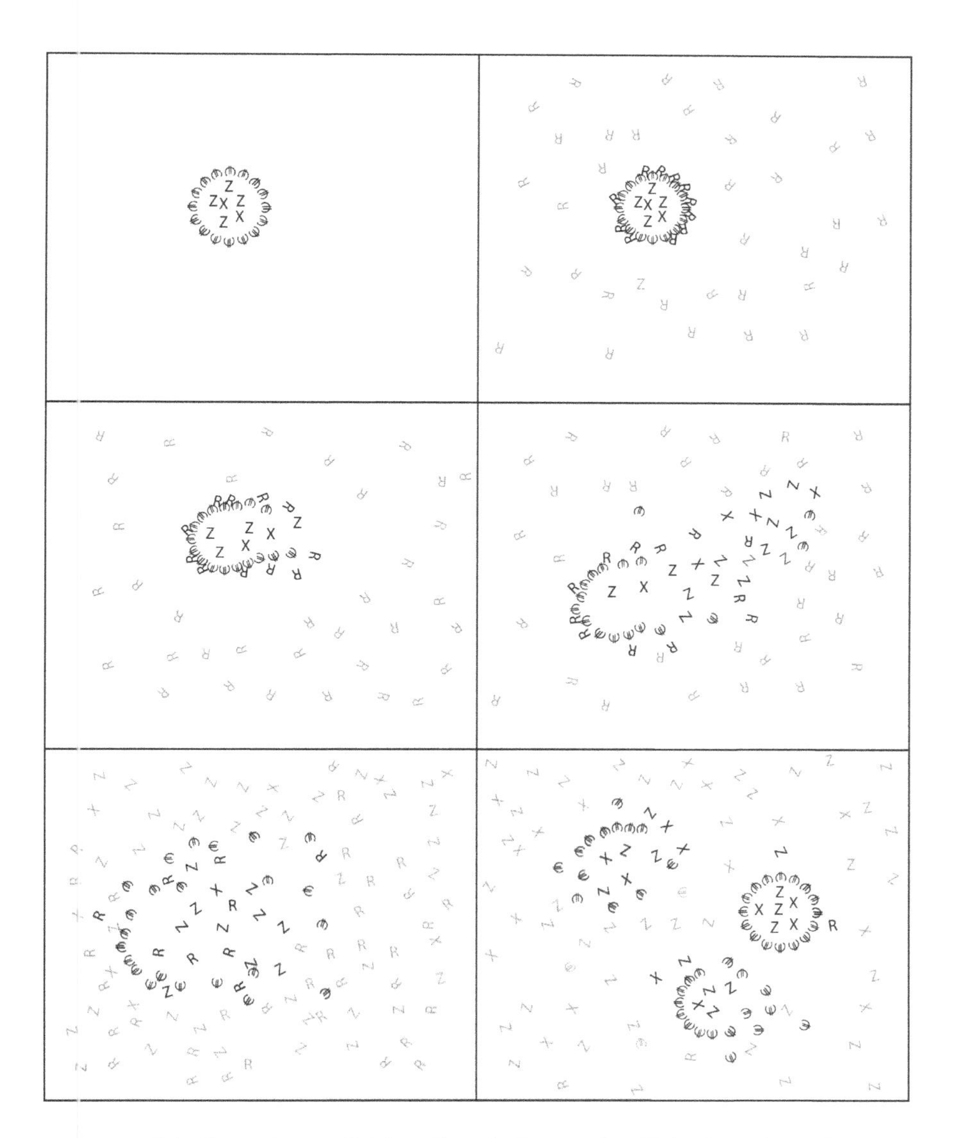

FIGURE 8 Selective autogen. Catalyst Z, replacing catalyst Y in the autocatalytic set, yields a variation on the capsid molecule by-product, (€), which tends to bind to reactants (Rs), yielding a rougher capsid surface, more prone to breaking. When capsids are broken, Rs float free, becoming available as reactants for the catalytic set.

The selective autogen would thus have evolved a modest capacity to interpret a difference in its environment, being more likely to open in the presence of reactants and otherwise stay closed.

Of course, this is one of many possible autogen lineage variations, most of which would be disadvantageous. The point here is that there's nothing about the selective autogen's evolution that defies physical law, and should it evolve, it would have both an adaptive advantage and, given self-regeneration, a capacity to "remember" the adaptive advantage, passing it on from generation to generation.

Since, in evolutionary theory, selection refers to the environment's effect on lineages, we must be careful to distinguish our use of the term here. Here, it is the autogen that does the selecting and does so by means of its evolved capacity to interpret.

Like the minimal autogen, the selective autogen is a thought experiment, a testable proof of concept, not a species found in the lab or in nature as yet. But its chemistry is nothing special or infeasible. The selective autogen demonstrates the kind of evolution that might be possible from a minimal autogen.

WISDOM TO KNOW THE DIFFERENCES FOR SELF-REGENERATION

The selective autogen's advantage reduces the likelihood of two potentially counterproductive scenarios: releasing catalysts in a reactant-free environment and not releasing them when there are reactants available to transform.

Obviously, the selective autogen doesn't intend to be more efficient than the minimal autogen. It just is, and therefore is more likely to continue to self-regenerate. Its simple ability to interpret circumstances is nonetheless a kind of know-how. Its capacity for selective interaction increases its self-regenerative efficiency.

With a little poetic license, we could even describe the adaptation as a kind of wisdom, consistent with the ever-popular serenity prayer: *grant me the serenity to accept what I can't change, the courage to change what I can, and the wisdom to know the difference.* The selective autogen has evolved meager wisdom to interpret a difference between environments in which it can and can't self-repair and self-reproduce.

This suggests a grounded, strictly pragmatic definition of wisdom as fittedness, the self's ability to interpret its environment efficiently and productively given its aims—most fundamentally, its aim to self-regenerate.

In distinguishing between the environments in which to open and close, the selective autogen reduces false positives and negatives on the implicit question "Open now?" A false positive would be opening where there are no reactants, doing work that is likely to fail to maintain self-regeneration, a sacrifice of protection with insufficient gain in repair and reproduction. A false negative would be staying closed when reactants are present: a missed opportunity to do work that is likely to succeed in self-regeneration, a sacrifice of an opportunity to engage in self-repair and self-reproduction.

The serenity prayer quest for wisdom to know the difference between what can and what can't change is likewise an attempt to reduce false positives and negatives on the question "Can I transform this to my advantage?" With the serenity prayer, the false negative is having the serenity to not try to transform what could be transformed to the self's advantage and the false positive is having the courage to try to transform what can't be transformed.

Obviously, the selective autogen isn't deciding or choosing whether to be open or closing. Still, we can see in even the minimal autogen a possible grounding for our understanding of choice even in humans.

Every day we make choices, most of them unconscious. At any given moment we are capable of doing myriad things that we wouldn't even consider doing. Choices are constrained away by evolution, instinct, habit, learning, and the constraint of circumstances that result from past decisions. The origin of choice is emergent self-regenerative constraint, limitations on what is likely to occur that emerges at the origin of selves and aims.

HOW A MOLECULE COULD BECOME ABOUT SOMETHING

The currently dominant origin-of-life theory is called the *RNA World* hypothesis. As I touched upon earlier, the idea that life is adequately characterized by replication of molecules—so-called replicator-selection theories—leads many origin-of-life researchers to assume that life probably started with DNA or RNA copying itself, perhaps with layers of clay facilitating the process.

This approach found further support when it was recently discovered that, under rare conditions, RNA can act as a catalyst and therefore could participate in autocatalysis, potentially facilitating its own replication. A recent *Nature Reviews Genetics* article even states, "this idea has gone from speculation to a prevailing idea."[1]

RNA World scenarios are problematic for two related reasons. First, though RNA plays an information-bearing role in all known organisms, it is not inherently informational. A strand of RNA drifting about in a prebiotic universe and therefore independent of a self isn't for or about anything, and it doesn't become for or about anything simply by participating in autocatalysis that facilitates its own chemical copying. The RNA World scenarios in all their variants are just an elaborate form of autocatalysis—chemical copying not for or about anything for any self.

Second, for RNA to become for or about anything the way it has in all known organisms requires mechanisms too elaborate to emerge spontaneously in prebiotic chemistry. An information-bearing role for RNA would require evolutionary steps and therefore evolvable selves existent prior to RNA playing any information-bearing role.

The minimal autogen and the selective autogen are models for evolvable selves emergent prior to RNA playing any role. Still, since RNA and DNA now play an information-bearing role critical for life on earth, it behooves us to explore how this role might have evolved. In other words, how might template molecules ever become about something for autogen selves?

Deacon offers a testable proof of principle scenario, a sketch for how autogens might have evolved toward the incorporation of information-bearing molecules. Here's Deacon's scenario in a nutshell, to be detailed somewhat more technically after.

A HYPOTHETICAL SCENARIO FOR TEMPLATES

DNA and RNA are polymers, linear chains of monomers. Organic chemistry, even in an abiotic universe, includes many kinds of other polymers, including sugars, carbohydrates, and proteins. Some polymers repeat the same monomers, and some link varied monomers, as do DNA and RNA, each with their four monomer varieties.

Polymers like DNA and RNA sequence the varieties as easily one way as another. You can think of it as like a string of varied beads that can be sequenced easily in any order. This achieves the aperiodicity that Schrodinger argued would be essential to an information-bearing molecule.

DNA and other polymers have an added feature. They can act as templates for making replicas of themselves, polymers with their four varieties in the same sequence. You can picture this as like a bead sequence—blue, red, green, yellow—that produces another strand with the same colors in the same order.

They do so indirectly, producing first an opposite strand that then yields the replica strand. To picture this, think of blue and red beads as reliably bonding to one another, and the same for green and yellow beads. The sequence blue-red-green-yellow would thus bond with beads in the order red-blue-yellow-green. This red-blue-yellow-green strand would then bond with beads that replicate the original strand sequence, blue-red-green-yellow.

This is roughly how polymer sequences replicate reliably—by pairing and unpairing with opposite sequences. You'll recognize a sequence-pairing as the DNA double helix. It looks something like twisted-pair wires or twisted bead strands. These strands would pair and unpair starting at their ends, like fraying wires.

RNA doesn't self-copy through pairing and unpairing the way DNA does. RNA World researchers assume that RNA is eventually replaced by DNA as life's primary information bearing molecule.

Textbook illustrations of double helixes show them as neatly symmetrical, but in reality they aren't. You can imagine this as like the colored beads being of slightly different shapes and sizes. Depending on the sequences, then, the outer surfaces of the double helix would have varied contours, like two strands of beads of varied sizes, twisted together. Other molecules can bond across the outer surface of a double helix, but only where they happen to conform to the varied contours.

None of this is purpose-driven chemistry. In it, there's no information about anything for any self. There simply happen to be molecules in organic chemistry that sequence flexibly, copy reliably, pair and unpair in double helixes, and afford a varied bonding surface that other molecules attach and detach.

Furthermore, DNA isn't the only polymer with this copying property. For example, sugars—energy-bearing molecules—can have these properties. An autogen that, in addition to catalyzing capsid molecule by-products, also

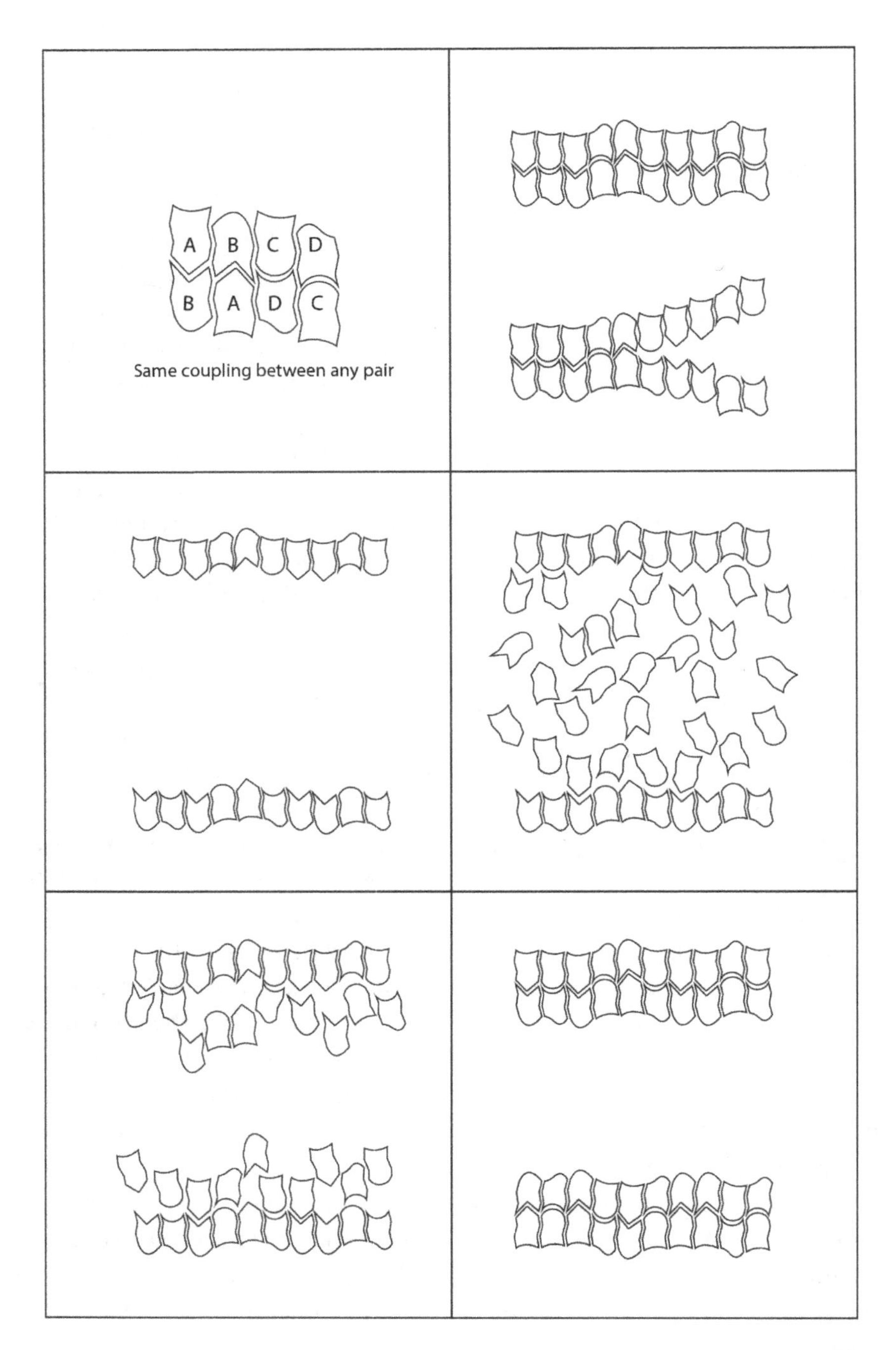

FIGURE 9 Toy model of template molecules: Molecule pairs (A/B and C/D) bond in any order, forming strands that split lengthwise, accumulating complementary molecules, thereby replicating original molecule pair sequences. This is a nonfunctional feature of some polymer molecules (for example, DNA, ATP).

catalyzed sugar monomer by-products would have an adaptive advantage, a capacity to store internally energy for autocatalytic reactions.

Deacon sees the bridge to an informational role for polymer molecules in the way that an autogen's catalysts might bond to the surfaces of sugar double helixes depending upon their contours. When the double helixes split into separate helixes, the catalysts would be released, not all at once but in a constrained order with catalysts near one another released at roughly the same time. Given the risk of error catastrophe, different constrained catalyst-release sequences would make a functional difference to the survivability of autogen lineages.

Long sugar polymers inside autogens would originate as random base-pair sequences of monomers. Like DNA or RNA, the base-pair sequences would, as a function of chemistry alone, divide, becoming two long single-base sequences. Free-floating bases would then bond to each single-base strand, resulting in two copies of the original base-pair sequences.

The sequences would be random because the sugars are only functioning as an internal source of energy facilitating autocatalytic reactions. In that functional role order makes no difference. Still, given the self-copying property of sugar monomers there would be heritability, different autogen lineages reliably inheriting different random monomer sequences.

Different random monomer sequences would result in varied polymer surfaces. Catalysts would bond to these surfaces based on this variability. When base-pair strands split, starting at their ends, catalysts would be released, starting at the strand's ends. Thus, random base-pair sequences would reliably yield different catalyst-release sequences. Catalyst-release sequences would constrain catalyst availability. Some random catalyst-release sequences would result in more efficient autocatalysis, thus resulting in some autogen lineages fairing better than others in reducing potential for error catastrophe.

Natural selection would eliminate autogen lineages with inefficient catalyst-release sequences. Surviving lineages would be those carrying forward heritable base-pair sequences that coded for efficient catalyst-release sequences.

What follows is a more technical description of how template autogens could evolve. If it is more chemistry than comes easily, one can skip ahead.

While many researchers focus on the fact that RNA can be both a catalyst and a template molecule, Deacon focuses on another curious biochemical coincidence. Versions of the same nucleotide molecules that constitute the skeleton of DNA and RNA molecules also serve to transport highly reactive

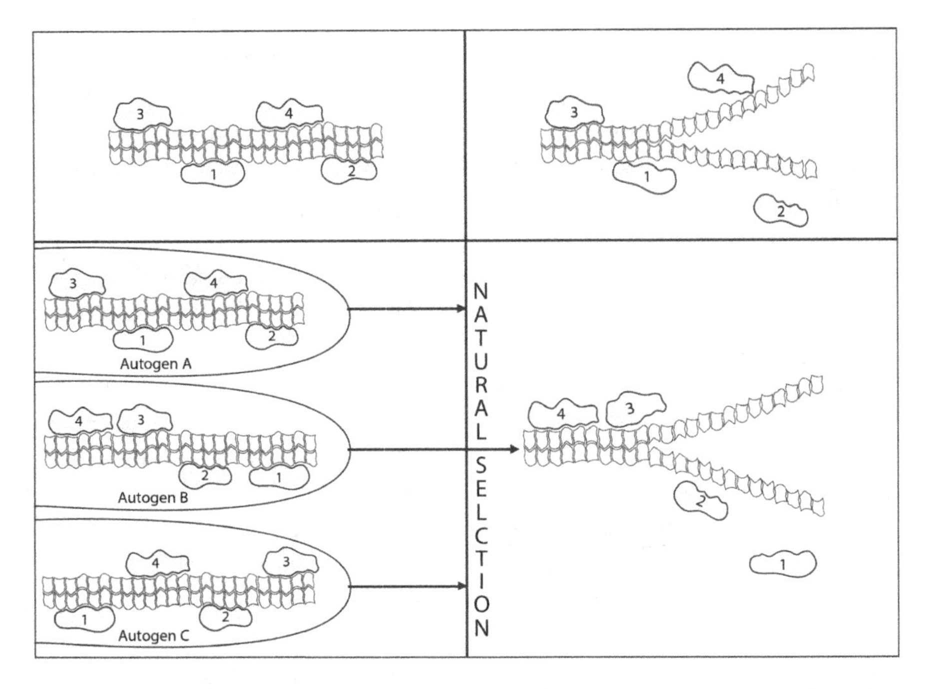

FIGURE 10 Template autogen selection. Catalysts 1 through 4 bind to contours of template surface depending on template sequence. When the template splits lengthwise, molecules are released in sequence. Templates that yield release sequences more conducive to preventing error catastrophe are selectively passed on to future template autogen generations. This suggests a way that a template molecule can become about something for an autogen self.

phosphate molecules (PO_4) from place to place within the cell, where they provide energy to aid a wide variety of molecular processes.

Phosphates provide a sort of common currency that fuels chemical reactions in all organisms. The major phosphate carrier is the monomer ATP (adenosine triphosphate). It consists of the same base (adenine), sugar (ribose), and phosphate that provide the "A" subunits of the A, G, T, and C molecules that constitute the aperiodic cross links in DNA.

ATP molecules are, in all modern organisms, like roving wallets, picking up, dispensing, and recycling phosphate energy currency. The average human body contains only about 8.8 ounces of these ATP wallets at any one time, but recycles the body's full weight in ATP every day. Without ATP, nearly all metabolic functions would come to a halt.

JUST ENOUGH BACKGROUND: ENERGY

As noted in part 4, whereas crystals give off energy when regularizing (for example, cooling a liquid removes heat, resulting in solids), autocatalysis typically requires energy. In minimal and selective autogens, the autocatalysis depends upon the energy extractable from reactants. This dependence on reactants severely limits the range of reactants and catalysts that could participate in the minimal and selective autogens' autocatalysis.

If there were a way that, in addition to capsids, autogens could also produce energy-ferrying molecules as by-products, then a wider range of reactants could be catalyzed by a wider range of catalysts, thereby making possible more autogen variation, some with larger autocatalytic sets than our X and Y example. This would be a significant adaptive advantage, though with more intricate autocatalysis, increasing the risk of error catastrophe.

If one or more of the autogen's catalytic processes produce side products that, like nucleotides, are capable of capturing and transferring phosphate molecules to energize other catalytic reactions, then one might expect this to provide a selective advantage. And a source of energized phosphate molecules might be provided near volcanic vents on the sea floor (an environment that many origin-of-life theorists consider relevant).

Although it is quite a leap without chemical justification to assume that variant nucleotide molecules like ATP and GTP can get produced in this way, the specific chemistry is not critical to the plausibility of the general argument about the transition from mere chemistry to template molecules. The properties that *do* matter are (1) the capacity of many similar but not identical monomers (single molecules like ATP and GTP) to link up at random with one another to form a polymer (a chainlike molecule) and (2) to exhibit variation in its three-dimensional shape along its length, dependent on the particular monomer sequence. The four nucleotides that constitute DNA molecules exhibit these properties. So we will simply assume that the energy-ferrying molecules are another variant on template molecules, in other words, nucleotide polymers.

THE TEMPLATE AUTOGEN

From our exploration of hypercycles earlier, you'll recall that increasing the size of an autocatalytic set tends to lead to error catastrophe. So a critical limitation on the evolvability of minimal and selective autogens is that as the network of catalytic interactions gets even slightly more complex the probability of useless and damaging cross-reactions increases rapidly and error catastrophe inevitably bars the evolution of increasing complexity.

All known organisms with their thousands of interacting molecule types prevent error catastrophe by means of regulatory mechanisms that constrain the way their many molecules interact with substrates and with one another. Constraining the probability of harmful interactions is the role that a template molecule plays in Deacon's *template autogen*, a hypothetical model—not real chemistry—that illustrates how a sequenced molecule might come to be about something for autogen selves.

Deacon imagines a two-phase evolution from either the minimal or the sensitive autogen that would result in a template autogen. In the first phase, in addition to producing capsid molecule by-products, a variant form of autogen also produces energy-ferrying nucleotide by-products that capture energy in the form of phosphates and transfer it to other catalytic processes during the process of autogen seed repair. Like the catalysts and capsid molecules produced in this process, many of the synthesized nucleotides would also likely get captured in the closed autogen. This could provide a store of energy-capturing molecules that, if released again when the autogen is damaged and open, could repeat this function. During an autogen's inert phase, however, the energizing capacity of activated phosphate molecules would be irrelevant and could be possibly disruptive. Using some of the unused phosphate energy to link nucleotides into a polymeric chain (as in DNA and RNA) could render this unproblematic by using the phosphate side chains to link nucleotides together.

In this way, a randomly sequenced polymer could serve as a sort of storage system to be depolymerized when the autogen is damaged and is again in need of energy. Just as some capsid molecules are likely to be reused from autogen generation to generation, some nucleotides would be recharged and re-encapsulated from generation to generation.

Continuing with the DNA analogy, Deacon imagines that during the inert phase the captured nucleotides would polymerize to form double helixes and that catalysts would tend to adhere to the helixes' surfaces in locations where helix shape and charge and catalyst shape and charge were roughly complementary. Since the twist of the helix is determined by the particular nucleotide bonding sequence, the relative position where a catalyst would adhere along the helical template is a function of that sequence.

Given the flexible ordering of nucleotides along a strand, the initial sequences would be produced at random. But since sequence order distorts the double helix's symmetry, these random sequences might constrain the order in which catalysts adhere to a double helix's surface.

This order can take on functional value if it plays a role in biasing the interactions between catalysts and between catalysts and their substrates. This can occur if relative position affects these interactions by influencing which catalysts are available when.

There are reasons to expect that inside the limited interior of an inert autogen seed (possibly aided by the exclusion of water and the high concentration of catalysts and template polymers) catalysts would tend to become bound to a template molecule. But when open to the environment (and exposed to higher water concentrations) these attachments are likely to weaken.

If the order in which they are arranged determines the relative timing of their release, then there will be a definite autogenic reconstructive advantage to nucleotide sequences that constrain catalyst availability so that error catastrophic cross-reactions are minimized. Thus relative proximity of catalysts on the template could play a role in the relative probability of their interaction.

Catalyst-release sequence would make a difference to an autogen's self-regenerative efficiency, and benefit self-regeneration. Those autogen lineages that contain nucleotide sequences that produce inefficient release sequences will tend to fall prey to error catastrophe and not self-regenerate. Those that carry forward a record of efficient nucleotide sequences that prevent error catastrophe would tend to proliferate.

With template replication there is thus the possibility of natural selection for nucleotide sequences that more effectively minimize harmful error catastrophe and maximize those most favorable to autogenic repair.

An initial population of autogens with a variety of randomly produced template sequences could thus evolve toward a population with ever more effective sequences, so long as the sequence itself is also reproduced with the reproduction of the autogen that contains it. But the replication of the template molecule would not be the source of its informative value, even though it facilitates a progressive convergence toward better representation of an autogen's optimal chemistry.

Unlike template-first approaches, making copies is not what makes the template sequence informative. Rather it is its inclusion as a synergistic part of the autogenic process that is most critical. In this scenario, genetic information is an adaptation that enhanced evolvability, but it is neither the basis for life nor its first cause.

Deacon's template autogen can help explain how some of the constraints that determine selfhood in this simplest sense can become offloaded onto the structure of a molecule. Of course many chemical steps are left unaddressed in this scenario, including the synthesis of nucleotide-like monomers, the mechanisms of template polymer formation and replication, and the conditions that determine the ways that catalysts become bound to and released from the template. These are daunting chemical issues. They may take decades to resolve and possibly in very different ways than suggested here—but the model suggests a possible bridge from pre-RNA World selves to selves that employ template molecules to constrain interaction, yielding a way to constrain the potential for error catastrophe.

23

WHERE IS THE SELF?

ARE AUTOGENS SELVES?

It would be hard for any of us to identify personally with the autogen. Its selfhood and aims are much simpler than ours, or even than what we find in any know organism. But to solve the mystery of purpose we must train ourselves to expect such simplicity. Highly intricate selves can't emerge from mere chemistry, a point overlooked or sidestepped by any researchers who assume that the first living organisms (selves) were already capable of writing and reading a record of their successes in DNA or RNA.

Still, we can't afford to confuse simple with impressionistic based on our subjective reactions, our eye-of-the-beholder interpretations of dynamics. As observers aiming to discover the origin of aims, our aims can all too readily contaminate our observations. It's too easy for any of us, scientists included, to project our aims onto things that don't have them. We're all born blurring the distinction between cause-and-effect phenomena and means-to-ends behavior, and our languages and cultures have tolerated the blur for millennia. With the autogen model, we must determine whether selves and aims are truly, not just impressionistically, present, even if they are inchoate.

We can expect attempts to produce autogens in labs within the next few years, with researchers working by trial and error to find the right chemical formula for generating autogenlike dynamics. That trial-and-error process would engender many errors, for example, minimal autogens that open or close too readily, or that, when open, autocatalyze too slowly or too quickly to be viable.

If researchers ever succeeded in discovering a means to produce autogens, they would step back and watch as autogens took on a life of their own, cycling through open and closed phases, proliferating autogens, and even perhaps evolving.

The researchers would be as excited as Orville and Wilbur upon seeing their first airplane take flight, or as Conway when his Game of Life produced animated patterns. Indeed, more excited, because a lab-generated autogen would be an empirical proof of concept for a solution to one of the biggest mysteries in science.

Or would it? The researchers might claim that they had discovered how life emerged spontaneously, but with all of their trial-and-error efforts, wouldn't this just be another case of *amnesic watchmaker syndrome*, researchers aiming to design and engineer selves and aims and then ignoring that they had done so?

Whether lab-generated autogens would be just another case of amnesic watchmaker syndrome depends on whether the autogenic chemistry that the researchers discovered by trial and error is of the kind that could have arisen spontaneously without the researchers' design and engineering efforts. A plane or a game as intricate as Conway's could not possibly emerge spontaneously in an aimless universe, but a minimal autogen is likely to be simple enough.

If it is, it doesn't matter that lab researchers aimed to produce it. So long as it could also emerge spontaneously, it would serve as empirical evidence for a plausible, strictly physical origin of selves and aims.

And even if the autogen could not arise spontaneously from prebiotic chemistry, the lab production of one would still be proof of concept for something heretofore not proven: that it is possible for means-to-ends behavior to be created from cause-and-effect phenomenon, something as yet undemonstrated despite decades of research in artificial intelligence, artificial life, and protocell research—research that attempts to build living cells from parts taken from existing cells. The point of the autogen theory is not fundamentally about the origin of life, but about the plausibility of self-regeneration emerging from aimless causality.

Still, if an autogen were simple enough to emerge spontaneously, what would distinguish the researchers' successful trial from their prior errors?

Not something added, but a reduction in possible paths emergent from the coupling of autocatalysis and capsid formation dynamics—a reduction that is our definition of synergy here, the whole produced by a coupling resulting in less than the sum of all possible dynamic paths.

RECIPROCAL MEANS TO PREVENT ENDING

Recall that philosopher Kant said that "The definition of an organic body is that it is a body, every part of which is there for the sake of the other (reciprocally as end, and at the same time, means)," but that he was unable to tell whether this was just his impression as an observer or a distinguishing natural characteristic of organic bodies.

"Reciprocal means and ends" isn't quite applicable to autogens since emergent regularization processes don't have ends. But they do end. Autocatalysis and capsid formation peter out. If either dynamic had an end or aim, which it doesn't, it would be to end—that is, to reach maximum entropy, succumbing to the second law tendency toward irregularity.

However, when these two emergent regularization dynamics are synergistically coupled, they prevent each other from ending. Thus, the synergistic coupling in autogens is *reciprocal means to the prevention of ending.*

With emergent self-regeneration, a true aim emerges, the aim of not ending. Obviously, it's not a felt aim, but it's an aim nonetheless, not one that observers project onto the dynamics but one that is evident from the fruits of its own constrained labors, the autogen's aimed resistance pitted against the second law. The self's synergistic coupling is a synergistic constraint or prevention. Each underlying emergent regularization process stops the other short of petering out.

Autocatalysis, by itself, peters out with catalysts dissipating. Capsid formation, by itself, peters out with capsid molecules dissipating. Synergistically coupled in the autogen, the autocatalysis and capsid formation dynamics prevent each other from petering out. Before all catalysts dissipate, some are contained in capsids, held together, thereby reducing the likelihood of catalysts just dissipating. Through the autogen's reciprocal constraint between underlying emergent regularization dynamics, paths toward catalyst dissipation become less likely to be taken. As a result, the remaining paths become more likely, increasing the likelihood of resuming autocatalysis when capsids break.

Likewise, as capsid molecules dissipate through second law degeneration, autocatalysis replenishes them by producing more capsid molecule by-products, thereby reducing the likelihood of capsid molecule depletion. Through this process of elimination of paths, paths toward dissipation of capsid molecules become less likely. As a result, the remaining paths become more likely—there is an increased likelihood of protecting and preserving autocatalytic sets.

If autocatalysis and capsid formation individually fall to energetic rest, irregularity, or maximum entropy, the synergistic coupling between them keeps them both aloft. Whereas emergent regularization results from dynamics falling to energetic rest, the synergistic coupling *regenerates* the constraints that keep them from falling all the way. Falling all the way is death. Staying alive is staying energetically aloft, never fully succumbing to the second law tendency toward irregularity.

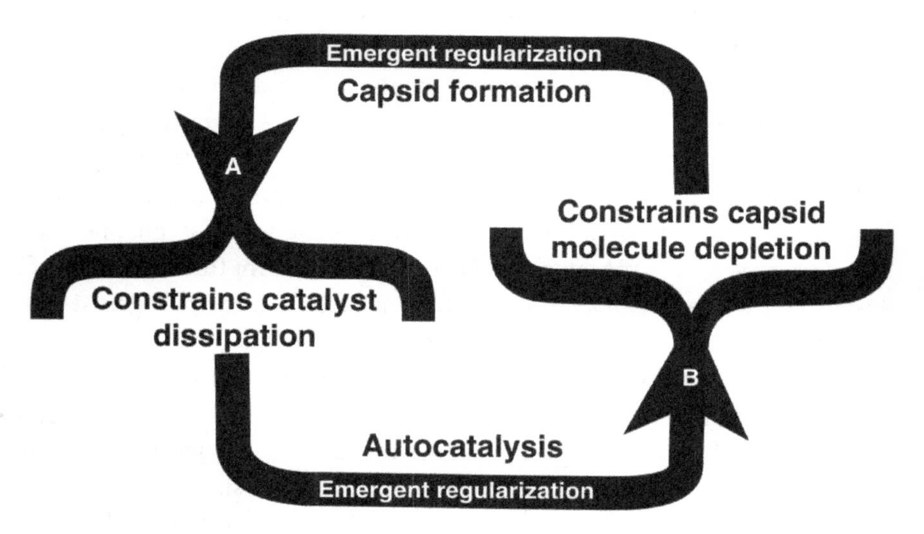

FIGURE 11 Schematic of the autogen's synergistic coupling: The two brackets represent reciprocal constraints that prevent underlying dynamics from degenerating toward irregularity. Capsid formation contains the autocatalytic molecules, preventing dissipation, and thereby permitting autocatalysis to resume if containments are broken in the presence of reactants. Reciprocally, autocatalysis prevents capsid molecule depletion by regenerating capsid molecule by-products. The autogen is the synergistic coupling that results from this reciprocal constraint dynamic.

Though we might assume that all selves are constantly energizing, for example, respiring continuously, this is too much too expect from first selves. To persevere, autogens have a nondynamic phase. When closed, they are inert like seeds, dormant, but with the potential to revive, returning, when capsids break, to dynamic interaction for self-repair and self-reproduction.

The autogen ratchets over time, cyclically building up (self-repair, self-reproduction) and locking down (self-protection) in dormant form. And how does it build up? Through interaction with external energy and resources.

USING THE SECOND LAW AGAINST ITSELF

Earlier I suggested that selves use the second law against itself. With the autogen model in hand we can see how this is so. As we saw with the pizza and oven example, energy is actually the second law tendency toward equalization or maximum entropy—segregated regularities (energy gradients, for example, the hot oven and cold pizza) interacting and becoming globally irregular. Energy is differences equalizing.

For self-repair and self-reproduction, the autogen relies on energy gradients. For example, in the minimal autogen, autocatalysis requires energy. Autocatalysis by itself is just equalization, the second law resulting in a regularization that puts up no resistance to the second law.

In contrast, autocatalysis within the autogen is a second law fall toward equalization stopped short of complete equalization. Through closure the autogen resists (self-protects against) the second law that enables it to resist. Whereas autocatalysis is just the second law playing out, autogens are the second law used to overcome the second law, thereby preventing their own termination.

Synergistic coupling makes self-regeneration's paths more likely by making other possible dynamics paths less likely, chiefly the paths that would lead to termination, as is otherwise the tendency of each of the two underlying emergent regularization dynamics. Emergent self-regeneration is the prevention of work toward dissipation that would otherwise occur if the underlying emergent regularizing dynamics were left to their own devices.

THE SELF IS NOT A WORKING MATERIAL OBJECT

People tend to assume that a constraint is a material thing. A riverbed is indeed a material constraint. But emergent regularizations are not. They are preventions that arise through material interaction but are not themselves material. A whirlpool is not a material thing distinct from the water molecules. Rather it's the way the molecules constrain one another in dynamic interaction.

We also tend to think of a constraint as doing work, for example, the work we do when we constrain ourselves in the face of temptation. But constraints don't do work. We saw this with emergent regularization. There's no internal attractor working to pull a whirlpool into existence, nor is there some holistic whirlpool form working top down to impose its spiral appearance.

Like emergent regularization, emergent self-regeneration is not a material thing and it does no work. It's a change in relative possibilities that result from interaction, lowering the relative probability of dynamic paths leading to degeneration and increasing the relative probability of dynamic paths conducive to self-regeneration.

It's hard to think of a self and its aims as an emergent constraint. We know selves by their material form and the unusual self-directed work that they do. We sure don't think of ourselves as the absence of possible dynamic paths. But this is just what I would argue we are.

It's easiest to appreciate in the difference between life and death. A body is material, but not the same material from day to day. We're not made of stuff so much as made *through* stuff, energy and material passing through. But at the moment of death the dead body is made of the same stuff as the live body just moments before, so what's the difference?

A live body is all of that matter and energy undergoing synergistically coupled emergent self-regeneration. The body's mutually constraining regularizations are the source of its emergent self-regeneration, a higher-level constraint emergent from the synergistic coupling of the body's lower-lever emergent regularization dynamics.

As with the autogen, the body's regularization processes constrain one another synergistically. At death, a body's synergistically coupled regularization processes lose their synergy and succumb to second law irregularity.

This is how the self is neither ghost nor machine. It's a synergistic coupling that achieves so much more by allowing so much less to occur than within a dead body. At death, the body's regularizing processes are no longer prevented from moving toward the anything-goes, anything-possible conditions of the second law. At that point all its chemical processes can run toward equilibrium unimpeded.

Synergistic coupling emerges with the first emergent selves. Their synergistic coupling emerges by accident. They are selves that were not generated by selves, aims that emerged aimlessly, and purpose that emerged purposelessly.

HOW A BODY'S SYNERGISTIC COUPLINGS ACCUMULATE

The minimal autogen results from a synergistic coupling of two regularizing dynamics: autocatalysis and capsid formation. "Coupling" could, of course, refer to conjoining just two dynamics, but it need not be restricted to two. Two is minimal and therefore the most likely synergistic coupling to emerge from abiotic chemistry. In evolved organisms a large number of regularization dynamics are coupled.

Synergistic couplings accumulate by evolution, as we saw with the selective and template autogens. Self-regeneration persists throughout a lineage while the underlying emergent regularizations can vary and evolve toward more efficient means to maintain self-regeneration. A living human body is the evolved accumulation of synergistic couplings involving myriad emergent regularizing dynamics that constrain one another.

The synergistic coupling between autocatalysis and capsid formation first occurs by fortuitous chance in an abiotic universe. Once that coupling emerges, so too does self-regeneration. With self-regeneration, selfhood persists over time. Evolution therefore becomes possible, tuning existing synergistic couplings, but it also makes it possible for additional synergistic couplings to accumulate, more dynamics interacting in tuned ways that maintain and enhance self-regenerative efficiency.

Synergistic coupling between individual organisms is called *mutualism*. There's a forerunner to this mutualism within individual organisms in what we commonly think of as a body's division of labor, synergy, or the codependence of a body's dynamics upon one another. This is the "all for one, one for all" tendency of a body, the formative power that Kant sought, whereby parts (actually, coupled dynamics) are "reciprocal means to ends" to one another's continuance.

We find synergistic coupling first in the minimal autogen. We see it expand through the selective and template autogen. With the template autogen we see a radical new source of further expansion, an enhancement for self-regeneration. The template provides a medium upon which features of the dynamics can begin to be offloaded, allowing for accumulation of intricate synergistic couplings without risk of error catastrophe.

It has been useful so far to distinguish four features: synergistic coupling, self-regeneration, selves, and aims. In the autogen they emerge as one and the same. The autogen is a self, but it is also the aim to continue self-regenerating, made possible by the synergistic coupling. What is this unity then that I have been at pains to explore?

It is a constraint that regenerates itself, a constraint that channels energy into work that sustains the constraint that channels energy into the selfsame work. The constraint is the means, the means whereby the end is achieved, the end being the maintenance of that constraint. Maintenance of that constraint is the end by which the means to maintain it is achieved. The self is not a thing apart, not a thing at all, not a ghost, and not the machinery. It is the self-regeneration of the mechanism, a constraint that enables the mechanism to continue to exist despite the second law.

So far, I have described emergent self-regeneration as yielding only three basic capacities, self-repair, self-protection, and self-reproduction, with the adjunct property selective interaction. The outcome of these three is the means to overcome the second law, which would be what selves would have to do first and foremost to exist and evolve as they have done here on Earth. Still with the emergence of self-regeneration there are many consequences we can now explore.

24

THE CONSEQUENCES OF SELF-REGENERATION

THE EMERGENCE OF TRYING

The mystery of purpose is also the mystery of trying: What is trying and how did it start in an aimless universe in which nothing is trying to do anything?

Trying often has psychological connotations. It conjures up humans trying because they feel the urge to achieve something or because they make a conscious commitment to trying. Still, it's commonplace to assume that bacteria are trying to stay alive. Bacteria don't have feelings or consciousness. Apparently feelings and consciousness are not essential to trying.

To try is to work against dynamic tendencies. Trying is therefore an inherently relative concept. There's what's likely to occur and then there's trying to have something else occur instead. A ball doesn't try to roll downhill. It rolls without trying. All selves try to stay alive. They're self-regenerative, regenerating their regularities against the second law tendency toward irregularity. It's as though everything in the universe is on a playground slide, sliding down toward irregularity, but selves are scrambling up it trying to keep regenerating their regularities.

Selves resist nonexistence. By contrast, nonliving things put up no resistance to nonexistence. There's no fight in them, no struggle against nonexistence. With nonliving things, more durable forms will last longer than less durable forms, but neither are trying to last. When researchers apply the concept of natural selection to all cases of relative durability, for example, treating galaxies or ideas as evolving through differential survival, they ignore the distinction between relative durability and trying to endure, or they employ trying metaphorically, for example, in Richard Dawkins's argument that memes or ideas are trying to survive and replicate.

Scientists have been unable to explain what trying is or how it got started at the origins of life. Oddly, most origin-of-life theories wholly ignore the emergence of trying. For example, in the most prominent theory today, life starts when RNA molecules replicate by autocatalysis. RNA is a chemical and chemicals aren't trying to do anything, even when they're replicating. Of course, in life today, RNA or DNA aren't just any chemicals. Scientists often describe them as instructions for building bodies. Still, while we might use instructions when we're trying to achieve something, the instructions aren't trying to achieve anything. A sterile planet brimming with replicating RNA molecules, whether you call them chemicals or instructions, would still have no selves on it trying to do anything.

Evolution by natural selection does not explain trying either. It is merely the name we've given to what happens when the living try to self-regenerate and only some of them succeed. Natural selection explains how living and trying evolve but not what living and trying are.

Many nonliving things can look from the outside like they're trying, a river trying to reach the sea, whirlpools trying to form, RNA molecules trying to replicate, computers trying to process information, so we need a clear way to distinguish what's really trying or we end up thinking that everything or nothing is trying, that rivers are alive, or that you're no more alive than a computer. The right distinction is self-regenerative constraint, the original and most fundamental kind of trying, the self trying to maintain its ability to keep trying.

Selves don't just appear from the outside as though they're trying. They try for their own sake. Selves constrain the flow of energy into work to try to prevent their otherwise inevitable degeneration. This most basic form of trying is what Deacon argues emerges with the autogen.

The term *trying* captures another a fundamental feature of life: the potential for failure. All selves try but not all succeed. Self-regenerative constraint is the source of the trying trials in evolution's trial-and-error process. Natural selection, the source of the error in the trial-and-error formulation, is not selecting or judging trials for natural selection's sake. The error in question is error only for selves, the entities that try, the beneficiaries of success and the victims of failure in the trial-and-error process. The self is both the thing that tries and the beneficiary of the trying. Trying is work originating in the self for the self's benefit, work against likely tendencies.

There's no trying in the first kind of emergent constraint, emergent regularization or self-organization. It can appear from the outside as though there's trying—whirlpools or Benard cells trying to form, autocatalysis trying to generate more catalysts, capsid molecules trying to form capsids. But there's neither work against tendencies nor a beneficiary for such work if there were any. Emergent regularization processes do not constrain energy into work against the tendency toward irregularity. They're not trying to stay regularized. Indeed, emergent regularization processes are ultimately degenerative in complete conformity with the second law. Whirlpools increase the flow of water; they don't try to resist it. Crystals form because their molecules are in a lower-energy state than if dissolved in a supersaturated solution. Autocatalysis too, including the autocatalysis of RNA molecules imagined in RNA world scenarios, is the product of potential energy being depleted in conformity with the second law, like a ball rolling downhill—no trying involved.

Trying emerges with the second kind of emergent constraint, the emergent self-regeneration we find in the autogen. It's easiest to see the autogen's trying in its capacity for self-repair and self-reproduction when the autogen seed gets broken. The autogen constrains energy into work that regenerates seeds. That's the autogen trying to scramble up the second law's slide toward irregularity. Obviously the autogen doesn't realize or feel like it's trying to work its way up the slide. Still, its constrained work against the second law is evidence of its trying.

Self-regenerative constraint is the synergistic constraint that results from the interaction between two nontrying processes, autocatalysis and capsid formation. These two processes aren't trying to work against each other. Still, in their interaction, new constraints are generated such that they reduce the likely tendencies of either one alone.

Individually, the autocatalysis and capsid formation processes would naturally degenerate. Synergistically coupled in the autogen, these two emergent regularizing processes prevent each other's degeneration. Given capsid formation, the autocatalytic chain reaction doesn't peter out before seeds form, seeds that make autocatalysis more likely to restart. And when capsid shells break and capsid molecules dissipate, autocatalysis is right there replenishing the supply of capsid molecules. The autogen exemplifies at the most basic scale the difference between not trying and trying. It reveals how a synergistic coupling between nontrying emergent constraint processes at one level can yield trying at a higher level.

To try is a verb, useful for focusing attention on the dynamic processes underlying all life. Some researchers argue that there is nothing but dynamic processes in the universe, and that therefore the abstract nouns we associate with life, for example, *self, aim, purpose, function, value, interpretation,* or *information,* aren't real. To these researchers, the universe is all verb. There are no nounlike abstract features that could distinguish life, because there's nothing but flux in the universe.

This assumption leads to eliminativism or panpsychism. Eliminativism argues that the nouns we apply to features unique to life are just abstract concepts that we impose on what is merely dynamic flux in order to make sense of it. Panpsychism argues that the nouns might as well apply to all dynamics, whether living or nonliving.

These theories don't hold up. They don't solve the mystery of purpose or of trying. They don't explain why living dynamics behave so differently from inanimate dynamics, or why selves try to counter the second law tendency toward irregularity when nothing else does.

The autogen approach suggests an alternative assumption about the relationship between dynamics described with verbs and the noun-abstracted qualities unique to life. When dynamics that try to maintain themselves emerge, features that transcend dynamics emerge also, the very real features that we recognize through our employment of nouns. The nouns identify real factors. They are not eliminativism's useful but unreal abstractions or the panpsychism's qualities applicable to any and all dynamics. The nouns identify real factors unique to life, persistent constraints that really limit and therefore alter what is likely to happen. Emergent with the autogen are the essential abstract features of life all rolled into one, and there are many such features, only some of them addressed in what follows.

The nouns point to neither material parts in the machinery of life nor ghostly immaterial objectlike nonobjects—vital force, soul, or spirit that animates the living. Rather they point to the constraining consequences of emergent self-regeneration.

With the autogen, we have a model for what a living, trying self is. It's not some ghostly force that enters a material body and drives it like a piece of heavy machinery, nor is it the machinery. A self is not the energy or matter that passes through the self or even the work the self does. Selves are self-regenerative constraints, constraints that channel energy into work to try to regenerate the constraints.

Self-regenerative constraint is also an abstract noun. Where do we locate it? Not in the matter or energy but in their dynamic interaction once the autogen emerges. The energy and matter come and go, but the self-regenerative constraint persists so long as the trying continues to succeed within a self and its lineage. The constraint exists in the change in likely dynamics, the possible dynamics prevented that yield the self presented.

THE EMERGENCE OF GOOD AND BAD

In a universe without selves, nothing is functional *for* anything and nothing matters to anyone *about* any circumstances. Once selves and aims emerge, events and dynamics become functional or not functional, in other words, good or bad for some self about its circumstances given its aims.

Good and *bad* are especially tricky words when seeking simple but not impressionistic interpretations of first selves. Based on our impressions, we may be rooting for an aimless dynamic, delighted if it persists, and disappointed if it doesn't. But that doesn't mean that something is good or bad for the dynamics themselves. For example, we can imagine that the dissipation of a whirlpool or autocatalysis is bad news for these two emergent regularizing dynamics. But the whirlpool doesn't care. Failure is not bad for the whirlpool itself.

Many a self dies without emitting a sigh of disappointment. So how can we tell what's good or bad, not from our perspective but from the perspective of the dynamics? By its consequences for the self-directed work.

The work that results from emergent regularizing dynamics can do lots of things. Tornados can spin and destroy large objects; autocatalysis can transform large quantities of reactants into products very quickly. Yet, nothing about the work that results from emergent regularization maintains the emergent regularization. A whirlpool is just a passive consequence of the second law. It puts up no resistance to it. The same cannot be said for self-regeneration. Self-regeneration *does* put up resistance to the second law.

Good and bad emerge with selves. What's good for the self-regenerating self is whichever constraints realize the potential to continue to self-regenerate. With the minimal autogen, there is for-ness and about-ness: behaviors are good for an autogen about or with respect to its circumstances.

The autogen doesn't know that the work that results from its constraints is good. It has no consciousness or feelings. The origin of life is not the origin of *like*. Some argue that it is, as though at their origin, selves exhibit responses we can describe as "yuk and yum," aversion and attraction.[1] To overlay those emotional responses on first selves goes too far toward anthropomorphizing impressionistically. Furthermore, yuk and yum are about taste, not self-regeneration. As anyone with a vice knows, there are plenty of yuks that are good and yums that are bad for our self-regeneration. Heroin, for example, is reported to be very yummy, though not conducive to self-regeneration.

Still, we can tell what is good and bad *for* autogens by what they do. Autogens channel work into self-repair, not self-dissipation. They channel work into self-protection, not indiscriminate interaction with their environment. They channel work into self-reproduction, not the termination of their lineages. Self-regeneration is good for maintaining self-regeneration.

The argument that good and bad emerge with life is consistent with evolutionary theory's focus on functional or adaptive traits that are good for organisms. Still, with no explanation for how selves and aims emerge, evolutionary theory has long been an incomplete explanation for the origin of good and bad.

THE EMERGENCE OF SELF-OTHER (NOT INSIDE-OUTSIDE)

It's easy to assume that all selves have a marked boundary between their material interior and exterior, a material container such as a cell membrane, skin, shell, or bark. Autogens do not have a constant containment. When an autogen is in its open phase, it is a local concentration of individual molecules, though one that has the dynamic tendency to reclose, returning to its seedlike phase.

Still, with autogens, there is a locus of self and there is a distinction between what is self and what is other than self. When an autogen is in its open state, it's not easy to tell which molecules will end up enclosed. Therefore, the self-other distinction is not identifiable by simple location. We can't always tell which molecules are part of and not part of the self.

But then, that's also the case for selves consistently contained within skin. The oxygen we inhale is not transformed into our energy with 100 percent efficiency. With every breath, there are oxygen molecules that enter our lungs, departing unabsorbed on our next exhale. Are these molecules part of our insides? The answer is not clear-cut.

Deacon would go so far as to say that a minimal autogen marks the first time we can describe something as a *system* not just in the eyes of the beholder. Researchers often describe dynamics as systems. They distinguish between closed and open systems, while recognizing that no system is entirely closed, isolated, and insulated from all outside interaction. For example, we can't bring a closed system down to absolute zero, the lowest possible temperature, because no matter how carefully it's insulated, there will still be some dynamic paths through the insulation between what is outside and what is inside the system. Therefore, the system is somewhat open. A closed system is an idealization.

For researchers, then, what distinguishes a system's insides from its outsides? The researcher's assessment of what is and isn't likely to be a substantial source of interaction. This is exemplified by Integrated Information Theory (IIT), as described in chapter 13, which assesses a conscious system as a function of relative interaction between bits of information (in the technical sense, interconnected bitlike switches switching one another). According to this theory, a relatively higher concentration of coupled bits constitutes a conscious system.

In general, researchers assess what is likely to interact with what, and then cordon off the higher concentration of possible interaction, identifying it as a "system" in determining the initial conditions from which to begin their analyses.

This is entirely appropriate. Deacon does not argue for some kind of holism wherein everything is equally connected to everything else. Still, he thinks it's important to ask when a system becomes distinct not just as a result of a researcher's guesses about what is likely to interact, but in reality.

Systems become ontologically (real-world) distinct when dynamics become a self, dynamics constrained such that they tend to maintain their distinctness, or to anthropomorphize the second law, the first time the second law gets any competition. Many objects and dynamics are slow succumbing to second law irregularity. Nothing besides selves channel work into regenerating regularities against the second law tendency.

There are researchers who argue that individual selves as distinct systems are not necessary at the origin of life. Chemist Addy Pross argues that dynamic chemical networks are sufficient to explain the origin of life. To Pross, "individuality is more a life strategy than a life characteristic. So-called individuality is just a technique that evolution has discovered, amongst many others, to enhance replicative ability and robustness."[2] Deacon, of course, disagrees, arguing that what must be explained is the distinct self-other relationship that must emerge at the origin of selves.

A self is an individual, not necessarily with a clear material border. Most fundamentally it is an individual due to its self-other antagonistic relationship with the second law, the way it, unlike any other dynamics, uses the second law against itself, regenerating regularities rather than just petering out.

THE EMERGENCE OF FORESIGHT

Ever since the Enlightenment, the elephant in the room, or rather in the science lab, has been telos, that is, end-directedness, purpose, aims, or final cause—that for which something occurs. Telos is inescapably about anticipating futures. It's about selves doing today what is likely to prove successful tomorrow in serving their aims. Means-to-ends behavior also implies the future: present means to future ends.

Scientists since the enlightenment have generally tried to escape telos. Evolutionary theory, explained by the law of effect, gave us our greatest hope of being able to ignore the elephant in the lab. In the standard interpretation of evolutionary theory, adaptive traits aren't for the future; they're just what remains from the past. According to this view, genetic variation has no foresight, nor does natural selection. Evolution is a passive contest with losers dying off along the way. Today's winners are what are left after the winnowing. Winning is not a testament to their future success, just to past success. Since the past is often prologue, there's a good chance that whatever survived until today will survive tomorrow too, but there's no guarantee.

Physical science research can be done with ample confidence that the elephant won't sully the lab. Cause-and-effect events clearly have no foresight.

They're not aiming for some future state. Through equivocation, it's possible to treat biology as having addressed and eliminated the elephant too. For example, through the term *teleonomy*, the mere appearance of telos makes the elephant an apparition, not something to tame or eliminate, but an illusion to ignore.

The social sciences don't succeed quite as well in avoiding references to telos, anticipation, and foresight. Humans plan. We clearly aim to achieve objectives tomorrow. We clearly anticipate future risks and try to avoid them. This may be one reason that the social sciences are called "soft" sciences. They are sciences in that they have long aimed to eliminate the elephant but just haven't figured out how to do it yet. Until they do, the social sciences, according to this perspective, are not pseudoscience, but a legitimate science still working on eliminating telos.

Deacon argues that we can't and shouldn't try to eliminate telos from the life and behavioral sciences. Telos is real. When a bacterium consumes sugar, it is anticipating, albeit unconsciously. It aims to acquire the nutrients today by which it will have the energy to keep aiming tomorrow. Its functional traits anticipate tomorrow's nutritional needs.

The autogen's capacity for anticipation is most intuitively recognizable in the selective autogen that is more likely to open when there are reactants present in its environment. Like the bacterium, it doesn't plan, yet its traits anticipate. It opens for future repair and the reproduction of future selves, the present means to future ends.

But even the minimal autogen anticipates. It already has self-regeneration fitted to what increases the likelihood that it can resist the second law and reproduce future offspring. Telos is a real physical phenomenon even in the minimal autogen.

THE EMERGENCE OF MEMORY

As conceived by Darwin, evolution by natural selection assumes the existence of memory, heritable traits passed from generation to generation. Neo-Darwinism identified the source of heritability as genes transferred from parent to offspring.

People describe materials as having memory also. A billiard ball compresses on impact but bounces back to its original form. Memory foam temporarily retains the shape of a body resting on it. In these examples, memory is a fine metaphor, but just a metaphor.

The term *memory* can also be applied strictly to psychological memory. By this standard, a bacterium remembers nothing, but we remember our childhoods. The word *memory* originally meant "recollection, awareness, consciousness."

When we think of memories as exclusively mental phenomena, they are among the mysteriously ghostlike features of mind. They arise only in the material mechanisms of consciousness, but are somehow different from them. Your memories have none of the features we associate with material objects.

The minimal autogen is certainly not conscious, nor does it have DNA that it passes from parent to offspring. Still, by means of self-repair and self-protection, the autogen "remembers" its regularizing dynamics over its existence. By means of self-reproduction it passes on its emergent self-regenerative constraints from generation to generation, literally re-membering or repopulating its environment with variations on its constraints.

In this, we gain a clue about where to look for all living memory, including yours about your childhood. We need not look for a material memory or assume that, without one, memories must remain as immaterial as ghosts. Memories are products of the constrained organization of dynamics. As such, memories are strictly natural but not material. They are functional constraints maintained through self-regeneration.

Let's now address a feature of what emerges with autogens that warrants a longer treatment, the transition from cause and effect to interpretation, and in the process explore the ways in which template molecules, such as DNA and RNA, come to be interpreted as information, and the ways in which such interpretation is like and unlike reading.

VI

THE INTERPRETING SELF

25

CODES, SIGNS, INTERPRETERS

FUNCTIONAL CONSTRAINTS

As we turn attention toward how selves interpret, let's zoom out to orient ourselves. Many cultural traditions regard the self as a soul that occupies a body transiently before passing on to or through the supernatural realm. The soul thus imagined is immaterial yet can do work—an immaterial cause of material effects. It migrates in and out of bodies, presumably disembodied in transit, a ghost whether inside or outside the machine.

In contrast, the self in the autogen model is a synergistic coupling capable of self-regeneration. Unlike the soul, it does no work. It is not an immaterial object that pushes and pulls, causing material effects. Rather, the self is the emergent consequence of material work occurring throughout dynamic interaction. Because it is emergent only through material interaction, the self is never disembodied. Even in self-reproduction there's material dynamic continuity from parent to offspring. In the autogen, the self is the aim to self-regenerate. It's an emergent constraint that yields function. It is an *emergent functional constraint.*

Starting with the minimal autogen as the model for the protoself provides us an opportunity to examine functional constraints starting from an uncluttered foundation, free from evolved distractions and philosophical abstractions that arise if we were to start higher up the tree of life. But what is a functional constraint?

With the minimal autogen, there are two obvious points where constraints play critical functional roles in maintaining self-regeneration. First, in the open autogen, we find a functional constraint in the way that, of all the times and places that autocatalysis and capsid formation could be occurring, they're constrained to cooccur in each other's midst, thus making it likely that capsids will come to contain a sampling of the catalysts being generated.

Of all the times and places that autocatalysis and capsid formation could occur, they are constrained to occur together. Autocatalysis produces capsid molecules constrained to just where and when they can function to regenerate encapsulated catalysts—autogen seeds. This functional constraint is temporal and spatial—the functionally right place, right time for reducing the likelihood of failure to self-regenerate. The spatial, temporal proximity is a functional trait though not an evolved adaptation since the minimal autogen, being a first self, has had no chance to evolve.

Second, in the closed autogen seed we find a concentrated sampling from the autocatalytic set. This too is a spatial and temporal functional constraint. Within the capsid the catalysts are dormant, but when the capsid breaks, the catalysts are released all at once in one location, which increases the likelihood that autocatalysis will resume. This is functionally advantageous because, though an individual catalyst might initiate autocatalysis, it is far less likely to do so than many catalysts released at the same time and place.

Thus, in the autogen we see two functional constraints, one dynamic (concurrent autocatalysis and capsid formation) and the other static (catalysts clustered together in the autogen seed). They are synergistic constraints in that they complement each other, each promoting the likelihood of the other recurring. They are functional in that they are conducive to self-regeneration. They are constraints in that they limit paths, reducing the likelihood of nonfunctional dynamics and thereby increasing the likelihood of functional ones. Both of these nonfunctional constraints depend upon spatial and temporal factors—molecules constrained to be in the right (functional) place at the right time.

Functional constraints, always embodied in material media, can be dynamic or static. Either way they are limitations on where molecules are likely to be that serve aims. The open autogen's functional constraint—autocatalysis and capsid formation occurring simultaneously in the same location—is an example of a dynamic functional constraint. The closed autogen's functional constraint—the sampling from the autocatalytic set clustered together within a capsid—is an example of a static functional constraint.

EQUIVOCATION ABOUT TEMPLATES, CODES, AND REPRESENTATION

The information-first approach to the origin of life focuses on replicating *template molecules* like DNA or RNA, treated as a static *code* or a blueprint that *represents* the traits that fit an organism to their environments. But the terms *template*, *code*, and *representation* are ambiguous, inviting equivocation. They refer either to cause-and-effect material phenomena or to functional means-to-ends behavior.

Consider the cavity exposed when a rock is somehow knocked out of the earth in which it was embedded. The cavity is shaped like the underside of the rock. We might say that the cavity codes for the rock. If mud settled into the cavity and hardened, we might say that the cavity was a template that yielded a representation of the rock's undersurface. These would be cause-and-effect uses of the terms.

In parallel, a catalyst could be thought of as serving as a template for the production of catalytic products in that, like the earth where the rock was removed, it has a concave bonding surface that molecules fall into, yielding a positive representation of the concavity. Yet catalysts don't serve some intrinsic function for selves. The interactions between catalysts, reactants, and products are strictly cause-and-effect phenomena.

In conventional use, however, codes, templates, and representations are typically functional for selves. When researchers confuse the distinction between functional and nonfunctional codes, templates, and representations, they make proposed solutions to the mystery of purpose easy to explain and understand, compelling, convincing, and wrong.

Researchers may claim that there is no mystery because the genetic code is just a template that represents adaptive traits programmed by natural selection. Researchers can claim that neuronal firings are a code that controls our behavior the same way that software coding controls computer behavior. They can sidestep the mystery of purpose by means of equivocation.

In attempts to solve the mystery we must always be careful to disambiguate all such equivocations. If we aren't careful, means-to-ends implications can be snuck into our analyses.

GENES ARE PATTERNS; DNA IS THE MOLECULAR MEDIUM

A molecule has no aims, but it can embody functional constraints. This is crucial to understanding the important yet sometimes overlooked distinction between DNA and genes.

Biologist George Williams, who inspired Dawkins's *Selfish Gene* approach, stressed the distinction when he said that "A gene is not a DNA molecule; it is the transcribable information coded by the molecule."[1] From the Williams quotation, we will have to unpack what it means to call the information "coded," and whether that's an apt description. But William's fundamental point is important: The medium is not the message. The message is a functional constraint within the medium.

DNA is a molecule; genes are the functional constraints embodied within these molecules. Biologist Richard Dawkins overlooks this distinction when he argues that "We are survival machines—robot vehicles blindly programmed to preserve the selfish molecules known as genes."[2]

Computer pioneer Norbert Wiener argued, "We are not stuff that abides but patterns that perpetuate themselves."[3] What he calls patterns I am here calling temporal and spatial functional constraints.

Given Dawkins's selfish gene paradigm, it would be easy to misinterpret Wiener as arguing that we (selves) are the patterns in DNA. Working from the autogen model suggests an alternative interpretation. We selves are patterns, both static and dynamic functional constraints, constraints that reduce possible paths. As patterns, selves are not material but are natural. Patterns exist. They only exist within material but they are not the material.

A pattern is a constrained spatial and/or temporal regularity within a material medium, constrained in that any pattern is one among many possible patterns. This is true of genes and also of text, which is one reason people often draw a parallel between the two.

The text you are reading here is a pattern, a constrained regularity, not just any letters in any order. It's a functional constraint. As this text's author, I have constrained letter configurations down to form a particular sequence of words conveyed by a pattern of pixels on my computer screen and a similar pattern within ink and paper in a printed book. However, the text is not the pixels and the screen or the ink and paper themselves. I have written the text to serve my functions and those of readers capable of interpreting it.

Like texts, genes are the spatial configurations, in other words, a sequence or pattern resulting from permutations of the four types of nucleotides A, T, G, and C arrayed on a strand of DNA. If DNA is the string of beads, genes are the bead sequences, functionally constrained to a pattern.

A pattern is not inherently functional. Genetic patterns, sequences, or constraints (they amount to the same thing) only become functional when they become about circumstances for selves.

Printed texts, genetic sequences, and catalysts clustered in a closed autogen seed are static patterns. Theoretical biologist Howard Pattee describes genes as "rate-independent,"[4] meaning that the constrained regularities maintained in genetic sequences are not altered by changes in energetic rates, for example, within a certain range, the amount of charge, heat, or pressure imposed upon the medium. He also means that their existence doesn't depend upon continuous energetic throughput. A genetic sequence doesn't fall toward irregularity when dormant. Nor does a book. In contrast, text on a computer screen disappears without energetic throughput. Rate independence is what I mean when I describe genes as static.

SIGNS

Like the functional spatial constraints in autogen seeds or genes in all known organisms, a text is a static functional constraint. It is constrained in that the letters are not in a random irregular configuration; and it's functionally constrained in that they serve writers' and readers' aims. Of course a material book—ink on paper—is no more a self than is the material molecules of an autogen or a strand of functional DNA. Nor is the constrained pattern of the text in the medium a self. Neither the medium nor the message is a self.

Rather text is a collection of *potential signs* for selves about circumstances. Signs are what people often refer to as information, for example, when we say that a stop sign contains information. Given casual, everyday materialist habits of thought, it's easy to leave it at that: information is a material object that causes behaviors, or information is the pattern in that material object. It's as easy as thinking that the oven contains a material called heat that we pump into frozen pizzas. It's also as inaccurate. Consider a stop sign:

A stop sign doesn't cause you to stop unless you crash into it: Information isn't a cause of effects. It's a sign that must be interpreted for it to make any difference.

A stop sign is not the metal and paint: That's the medium. The stop sign is the functional constraint or pattern within and upon the medium. Of all the colors, sizes, shapes, letters, and letter orders possible, it is in a somewhat constrained variety, red, hexagonal, and so on.

A stop sign doesn't aim to make us stop: It doesn't have aims of its own.

A stop sign is a potential sign: The sign is not functional on its own. A potential sign only becomes an interpreted sign if and when a self interprets it as such, doing things in response to it that they wouldn't do in its absence.

A stop sign is open to interpretation: It can mean different things to different selves depending upon their interpretations of it.

A stop sign is a member of a class of signs: It is one of a diverse variety of street or traffic signs. When selves interpret it, they first recognize that it's a member of that class.

Let's now look at the parallels to text:

A text doesn't cause concepts to come to mind: Text isn't a cause of effects. For texts to make any difference they must be interpreted by and for selves, about their circumstances.

A text is not the ink and paper or the pixels on the screen: Either of them is a medium. A text is functional constraints within the medium. Of all letters, letter orders, and configurations, the text is in a highly constrained configuration.

A text doesn't aim to influence us: It doesn't have aims of its own.

Texts are potential signs: Text is not inherently functional. It's only functional when a self interprets it as such, doing things in response to it that they wouldn't do otherwise.

A text is open to interpretation: It can mean different things to different selves depending upon their interpretations of it.

A text is a member of a class of signs: When you interpret a text, you have to first recognize that it is a text in a particular language, a class or system of signs that you have the ability to interpret, in this case English.

On this last point, when you look at this text you first recognize that its letters and their configurations are signs within a class of signs, a *sign system*. You don't interpret them as spilled ink, hieroglyphs, pictures, pointers, or words in a foreign tongue. You recognize this text as sequences of English letters, words, and sentences.

With written text, spatial sequence matters. D-o-g brings to mind the concept of a dog. O-g-d suggests nonfunctionality or perhaps a typo. "The dog is in the den" makes sense that "den in is the dog" doesn't. Spatial sequence matters.

With spoken text, spatial sequences translate into temporal sequences. The wrong spatiotemporal sequences can't be successfully interpreted as about anything. They fail to have reference to anything except perhaps careless communication.

PROMISCUITY

Do words code for concepts? No, they don't generate a causal, one-to-one mapping. The word *dog* doesn't necessarily bring to mind a particular canine. Rather it constrains concepts down through a process of elimination, narrowing as broadly or tightly as serves interpreters aims given their circumstances.

Words are *promiscuous*, which, as a word, is itself a good example of this quality. You can tell from the context that I don't mean sexual promiscuity, but rather that words can "hook up" to convey many concepts. The word *promiscuity* narrows in on a range of concepts on several dimensions, one of which is ambiguity or unfaithfulness, the opposite of unambiguous, faithful one-to-one coding.

haha

Promiscuity gives language its extraordinary flexibility both for creativity and for equivocation. With languages, we can communicate and interpret to get a gist, or to narrow in to as much precision as possible.

We are careful in our choice of words, aiming or narrowing down to refer to the range of concepts that serve our aims, which can be broad or narrow depending on our aims. Lawyers, for example, write contracts aimed for narrow interpretation, tightly constraining the concepts that their words bring to mind in order to prevent equivocation. Poets and con artists may seek equivocation.

Still, copying words from one medium to another is indeed a one-to-one coding process involving no interpretation. This is another important but often-overlooked distinction. Copying is cause-and-effect material coding. Reliable—not promiscuous at all. There may be variation, but there are ways to minimize the variation down to little differences that don't make a difference. We should not mistake causal coded copying for the interpretation of signs. When you read text, you are not copying signs from text to brain. When you transfer a document from your computer screen to print, the printer is not interpreting. With this in mind, let's dip back into the genetic "code" to explore the ways that it is and isn't a code.

CODING VS. INTERPRETATION

DNA causally codes for amino acids. Different gene sequences constrain amino acids that are linked together to produce proteins, the fundamental building blocks in bodies. Given the cell's mechanisms for transcribing DNA, a genetic sequence has a one-to-one correspondence to an amino acid. DNA sequences are like strings of nucleotide beads. Proteins are like strings of amino acid beads. The cell's transcription mechanisms generate a direct correspondence between three-nucleotide sequences (codons) and amino acids. For example, the codon sequence TGG constrains twenty possible amino acids down to one amino acid, tryptophan, and not at all promiscuously. There's no interpretation in the process of transcribing from codons to amino acids and therefore from genes to proteins. Constraints or patterns in one are transcribed directly and reliably into corresponding constraints or patterns in the other.

But that doesn't mean that genes are a code. Where genes become an interpretation is in the way they mediate the self's relationship to its circumstances. Genes are interpreted for selves about circumstances. How genes get their about-ness for selves is what we'll now explore, again using the simplest prototype. With the template autogen we'll explore the about-ness relationship uncluttered by complications.

You'll recall that in the template autogen model we imagined that a sequenced molecule bearing phosphates—an energy source—might have come to play an energetic role within autogens, encapsulated and energetically dormant as double helixes with catalysts bound to their surfaces within closed capsids. When the capsid breaks, these molecules might serve as an energy source facilitating autocatalysis. Stored in the seeds as double helixes, the sequenced molecules would be in any random, unconstrained pattern. Different random sequences would result in different contours to double helix surfaces.

Within the seeds the catalysts would bond to the surface of the double helixes. There would be a coded one-to-one correspondence between the random spatial sequence and the spatial sequence of the catalysts bonded to their surfaces. When the capsid breaks, the double helix unravels, starting at its ends like twisted-paired wires untwisting. As it untwists the catalysts break free of the surface in a temporal sequence.

Thus there is a coded cause-and-effect relationship between spatial template strand sequences and temporal catalyst-release sequences. At first the template strand sequences are strictly random since the template molecules are only serving an energetic function to which sequence makes no difference. Still, given the replication properties of double helixes, the random sequences would be heritable, meaning that catalyst-release sequences would also be heritable.

Since catalyst-release sequences make a difference to the likelihood of achieving self-regeneration, through natural selection, the template molecule sequences would eventually evolve toward nonrandom sequences coding for functional catalyst-release sequences.

This loosely parallels the correspondence between heritable genetic codons coding for amino acids. Sequences on the template autogen's template and in DNA copy reliably within their mediums—template molecules to catalyst-release sequences and genes to amino acid sequences in proteins. As such, they are both coded constraints—constrained, in that of all of the possible sequences they are constrained to their particular sequences. But this coding is not inherently functional unless a self interprets it.

Within the template autogen, the constrained sequence become functional when a catalyst-release sequence makes a difference to the efficiency of autocatalysis involving larger catalytic sets with risk of error catastrophe. This provides for enhanced productivity in the self's aim to sustain self-regeneration. As such, the template is a primitive kind of "wisdom to know the differences" that make a difference to self-regeneration, as is the selective autogen's tendency for seeds to break open more readily in the presence of reactants.

With the template autogen, we can thus make distinctions useful in thinking about whether DNA or genes are codes. Nucleotides aren't codes, they're mediums. Sequences within a nucleotide medium are codes in their ability to replicate reliably and copy to other mediums, in the template autogen from nucleotide to catalyst-release sequences, and in genes from nucleotide sequences to amino acid sequences within proteins.

They are codes in that they generate one-to-one correspondences. Still, they are not inherently codes about anything until they become functional for selves. They only become functional for selves when they make a difference to a self's ability to self-regenerate. Thus though genetic transcription is a coding process, genes only become functional when interpreted functionally about circumstances by selves.

Earlier I touched on current concern about computers taking over the roles played by humans. Our distinction between codes and interpretation is relevant to predicting the extent and limitations on a potential computer takeover of human roles. Anything that can be managed by coded algorithms is likely to be. Humans can offload a surprisingly large amount of what we do onto automated systems whereby we translate inputs into outputs reliably by means of computer coding. A lot of what we have relied upon interpretation to achieve can in fact be achieved through automation, but the takeover will never be a full substitute for interpretation. Selves interpret; computers and other automated devices are still a far cry from a capacity to interpret.

There's no fundamental physical constraint that prevents interpretation from being achieved eventually *in silico*. Perhaps someday computers will become interpreting selves even though no computer today is even close. Until there are hardware selves, we can rest assured that there are some jobs that only living selves can do.

Not that this should be a source of much comfort. Though there is concern that computers will come to life and take over the world, destroying what we find valuable, machines are at least as dangerous. Mindless machines misprogrammed by humans already expose us to extraordinary perils—accidental nuclear war, for example.

Let us turn now to exploration for a clearer explanation of what it is about selves that does the interpretation of codes.

WHAT INTERPRETS?

People tend to think of interpretation in psychological terms, sensing and responding consciously to serve our aims. Here I use the term *interpretation* more broadly. An interpretation is a self's embodied bet about what will achieve its aims in its circumstances.

As touched on earlier, all selves interpret their environments. Even the minimal autogen is a representation of what might work given its circumstances, for example, that the reactants necessary for autocatalysis can become depleted, that there is potential for capsids to break open, and, overall, the need to overcome and protect against the second law tendency toward ending.

The minimal autogen interprets its environment directly by means of its constrained dynamic paths. Thus, the autogen is an interpretation, a bet about what might work given or about its environment.

With the minimal autogen, we can't distinguish between sign and self. The self interprets through its overall emergent self-regularizing constraints. With the selective autogen, we can distinguish one sign relationship. A reactant attaching to the closed capsid surface is an interpreted sign to open, interpreted by the whole self but through the distinct characteristics of the capsid surface.

The surface becomes something like a simple sense organ, breaking open on contact with reactants and otherwise staying closed. Is it the surface that interprets contact with reactants as a sign? We might say so, just as we might say that eyes sense and interpret images.

But really, it's the whole selective autogen self that does the interpretation. The selective autogen is a bet, an interpretation on what will achieve its aim to self-regenerate given its circumstances, not an interpretive bet placed by natural selection to achieve its aims, not an interpretive bet placed by individual molecules or constrained molecular sequences, and not an interpretive bet placed by individual traits. Selves are both the interpretations and the interpreters.

With template autogens, signs become internalized as a distinct static part of the self. Whereas the selective autogen interprets reactants that attach to capsid surfaces, the template autogen interprets its internal signs. The template sequence is a record of past success in the form of functionally constrained sequences. And each possible sequence is a sign within a system of signs, much like this text is a sequence of signs in a language system, a parallel we will touch upon next because the same would be true of all known selves interpreting DNA as a heritable record of past successes in the genetic system of signs.

The self is the means by which the interpretation occurs. The self as an overall bet about what fits its circumstances. Obviously this is a far cry from the interpretation that we human selves do, to which we'll now turn, though with relevance to a remaining question: whether genes function the way languages do, given that it is a popular impression that they do—genes as the language of life.

26

KINDS OF SIGNS

ICONS, INDEXES, SYMBOLS

Languages and genes both involve variable constrained sequences of signs, with languages, letters, or sounds, with genes, nucleotides. Does this mean that genes are a language?

To answer this question, I'll next explore the distinction between symbols—letters, for example—and other kinds of signs. This brief tour of kinds of signs will also give us perspective on human processes of interpretation and our unusual selfhood. The following is based primarily on Deacon's early work in semiotics—the science of sign interpretation—that resulted in his first book, *The Symbolic Species*, and that will be further developed in his forthcoming book on information. Deacon's approach to semiotics is rooted in the theories developed by philosopher Charles Sanders Peirce, who coined the terms *semiotics* and *pragmatism*.

First, we should remember the distinction between potential signs and interpreted signs; absolutely any distinguishable state—a "bit," of any phenomenon—is a potential sign. It only becomes a real sign when a self interprets it as about something. Recall that a fifty-fifty chance of rain is a binary bit of potential information. With no self present to interpret it as about anything, it's just a physical phenomenon, merely a potential sign. Interpretation is how significance happens—signs for a self about something.

Languages comprise symbols, one of three kinds of signs distinguishable by how selves interpret them as being about something.[1] The three kinds of signs are (1) icons or likenesses, (2) indexes or pointers, correlated with what they're about by a temporal or spatial proximity, and (3) symbols—signs that are about something by social convention. While some researchers treat these distinctions as intrinsic to the signs, they are not. Rather, the kinds of signs are distinguished by modes of interpretation.

In English, the word *dog* isn't a likeness to a dog so it's not iconic of a dog. It doesn't point to dogs so it's not an index, either. The word *dog* is symbolic, representing dogs by social convention within the English-language system.

Depending upon a self's mode of interpretation, the same potential sign can be interpreted in more than one way. For example, a wedding ring can be interpreted as an icon, index, or symbol. The ring is iconic when it is interpreted as a likeness to a marriage, for example, as an unbroken circle representing continuous commitment, or as an unbroken corral that keeps the married couple united.

The ring can also be interpreted as an index, for example, because the ring was given immediately after wedding vows were taken and is thus temporally proximate and therefore pointing to wedding vows.

Or the ring can be interpreted as symbolic of marriage, for example, in a culture that employs a variety of clothing and accessories as a system of signs established by social convention, ties for business people, stripes for military officers, wedding rings for married people. It's all in how one interprets the ring. The mode of interpretation is not intrinsic to the potential sign.

BUILDING ICONS INTO INDEXES INTO SYMBOLS

It is possible to use the three kinds of signs as a simple taxonomy. But early in his research career, Deacon noticed how icons build toward indexes, which build toward symbols.[2]

By indexical pointing, we mean some kind of correlation in time and space. Pointers to a dog's presence would include seeing dog fur on a couch, hearing a distant dog barking, pointing at a dog, or seeing a dog's tail wagging around the corner, factors that, through repeated experience, we have learned accompany a dog's presence and make us expect one.

Pavlov trained a dog to expect food when a bell rang. Given the dog's interpretation, the bell ringing pointed to food coming. To the dog, the bell became an indexical sign of food.

The dog's interpretation is a little like "putting two and two together," multiple instances of iconic likenesses of two types converging toward an expectation. For example, Pavlov's dog had a series of similar experiences of the bell-food combination. Each rung bell was iconic of the others, each food presentation was iconic of the others, and each bell-food combination was iconic of the others.

Thus, Pavlov's dog experienced two iconic relationships, similar bells and similar food, within a third iconic relationship, similar bell-food experiences. The dog puts two or more bell experiences together, two or more food experiences together, and two or more bell-food experiences together. As a result, when the bell was presented without food, the dog expects food. The bell has become indexical, pointing to food.

Still, "putting two and two together" isn't quite accurate in that the dog isn't actively putting anything together. Rather, the dog is failing to distinguish between similar experiences.

Interpreting instances as iconic of one another is passive indifference to differences. The dog doesn't think, "Wow, this bell sounds like the last bell!" The dog simply fails to notice the difference. Camouflage works the same way. We don't say, "Wow, that camouflage looks like its surroundings!" We simply fail to notice a difference.

We can, of course, interpret iconic signs more actively, as when we say, "Wow, that picture looks a lot like her!" But at its core, iconism is obliviousness, an interpretive indifference to differences, an inability to distinguish between different instances of similar signs. When someone returns who has been out of the room for a minute, you don't say, "Wow, you look a lot like the person who was in here a minute ago." Still the person's second coming is iconic of prior arrivals.

In parallel to building up from icons to indexes, the interpretation of symbols builds up from indexes. A child hears the word *dog* when dogs are around. The child doesn't differentiate among the various ways that the word is uttered, or among the dogs that appear. Eventually, the child expects the word and dog to come together. The child shouts "dog!" when one appears, and the word by itself signifies dogs.

We call this "making an association," but like "putting two and two together," it's not quite accurate. The child doesn't have to actively create an association between word and object. The word *dog* and the presence of dog are iconically repeated experiences until they aren't. The absence of one or the other half of that otherwise repeated combination points indexically toward what's missing from the child's *dog*/dog experience.

WAKING UP TO SYMBOL SYSTEMS

Children don't have to actively make associations, but they do have to discover that symbols are part of a sign system, a language. Until that aha moment, the child may be able to treat words the way Pavlov's dog treated bells, not as sign within a system of symbols but as an index or pointer. From the aha moment, the child recognizes a systematic correspondence between configurations within a system of symbols and their correspondence to a system of concepts.

Consider the Helen Keller story. Blind and deaf, she had no way to know that there were symbols for concepts. It took active repetitive work by her tutor, Anne Sullivan, to get Keller to recognize that the tactile sign language Sullivan was teaching corresponded to concepts. Keller's famous aha came when she recognized that the symbol for water was about the water coming out of the water pump in her yard. Suddenly she intuited that the sign-language word for water was part of a *system* of symbols. As a result, she raced around touching things and gesturing that she needed to know the sign-language symbol for them. With that recognition, Keller found a bridge to her family and society, a bridge made by a system of symbols maintained by social convention.

Before her epiphany, Keller might have been able to handle these sign-language symbols indexically, interpreting independent pairings of sign-language symbols with objects or concepts much the way Pavlov's dog treated the independent pairing of bells and food. For Pavlov's dog, there's no system of signs. At least in Pavlov's basic experiment, the dog doesn't learn that one bell means food and another bell means walk or drink. Keller's aha moment happened when she realized that the whole sign-language system can represent the whole system of objects and concepts.

Most of us have this aha intuition early and unceremoniously, but it does open for us the whole world of symbolic systems. Given our evolved human symbolic competency, we end up with a vast and interconnected system of symbols that parallels the vast system of experiences of objects and ideas.

With symbols, then, it's important to note that they are tokens of a certain type, signs within systems of signs. When you look at a text, you instantly recognize that the squiggles and lines are words in a particular language, a system of symbols, English, for example, not hieroglyphic icons or indexical pointers, but also not Chinese, Russian, or mathematical formulas, each of which are distinct systems of symbols.

In humans, the interpretation of symbols takes on a life of its own, generating our mental models of reality and making it possible for us to imagine the past, present, and future, the possible and the impossible.

SO ARE GENES A LANGUAGE?

Genes are a system, but not a system of symbols. Their correspondences to amino acids are certainly functional, but they are not generated by social convention. Rather they are a product of evolution. In the autogen paradigm, they evolve from the minimal autogen, which protects its regularized constraints in static form as the seed, not in nucleotide sequences. We could describe genes as a protosymbolic system, the first system of tokens of a type. And there's a neat correspondence between our three kinds of autogens, minimal, selective, and template, and the three kinds of protosigns, which I'll explore very briefly next.

Deacon argues that in the three kinds of autogens we see protosigns that parallel icons, indexes, and symbols. Minimal autogens are inherently protoiconic through their self-regenerative sameness over time and lineages.

Selective autogens interpret the world protoindexically, since reactants binding to their capsid surfaces increase the likelihood of an autogen breaking open in the presence of autocatalyzable reactants. The selective autogen thus gains a protoindexical interpretive competency. The reactants that make it more likely that capsids will break open are interpreted by the selective autogen "pointing" to the presence of reactants that can be converted to products through autocatalysis.

Still, the selective autogen's opening in the presence of reactants is not a case of a token of a type or system of signs any more than bell-ringing was for Pavlov's dog. The selective autogen does not distinguish, for example, between different kinds of reactants, opening differently for each kind, although such an extended repertoire might evolve over time.

Template autogens interpret protosymbolically, because a system of nucleotide sequences has become adaptively correlated to a system of catalyst release sequences.

TEMPLATE ADVANTAGES

With the template autogen, there's a loose evolutionary parallel to the aha moment we each have in childhood, or more importantly the aha moment that happened when humankind discovered the power of language. The template autogen's protosymbolic competency opens new avenues for evolution, well appreciated and understood in biology.

The template sequences provide a static representation of a key aspect of the autogen's dynamics. They become a mechanism for constraining error catastrophe, a means for orchestrating catalyst release. Having this heritable advantage provides autogens with a static material medium upon which other functional constraints might accumulate.

In Deacon's model, the template thus provides a redundant medium for protecting and reproducing functional constraints. The template is redundant in that it re-presents the autogen's dynamics. Thus the template autogen carries forward two copies of the same constraints, one in the autogen's dynamic alternation between open and closed phases, the other in the static template molecule sequences.

You'll recall that Shannon recognized the ways redundancy makes for reliable communication, in our example the way that redundantly repeating signal will stay constant while noise varies ("I'll . . . there," ". . . be there," "I'll be. . . .").

In the template autogen, the benefit of redundancy does not serve communication but repair and reproduction, functional constraints passed on from generation to generation. Should an autogen's dynamics falter in yielding self-regeneration within one generation, the static, redundant, functional constraints carried in the template molecules might still enable the lineage to continue.

Having the static template might also enable greater dynamic exploration of variation, chance modifications to the dynamics that could stumble upon new adaptations that improve the efficiency of self-regeneration given the risk of error catastrophe, adaptive constraints that might also come to be redundantly represented both in the dynamics and in the template sequences.

Given the advantages of a redundant record of functional constraints carried forward in gene sequences, it's no wonder that many biologists have long assumed that genes are essential to life. The redundancy affords both greater heritability and variation without compromising reproductive success.

The problem with assuming genes are essential to life is that interpreting template molecule sequences from a cold start in a prebiotic universe is highly unlikely given the intricate mechanisms involved. The minimal autogen provides a platform—self-regeneration from which template molecules' intricacies could evolve as means to more efficient self-regeneration.

The template autogen's proto-aha loosely parallels what language has done for us, and especially written language, a static medium, text as an aperiodic crystal that has spawned the combinatorial explosion of concepts in human culture, a veritable brainchild population explosion of explored and recorded concepts over the past few thousand years of human cultural history.

NO INTERPRETATION WITHOUT SELVES

There have been many careful attempts to understand what information is and how it works. But theories of information will remain incomplete without a solution to the mystery of purpose. There is only potential information in a sign or signal medium unless and until it is interpreted. Interpretation is the process of determining that some sign is *about* something else *for* some self's aims.

Some researchers have simply assumed the existence of selves, as Shannon did with his assumed sender and receiver of communications. That made perfect sense, since his focus was on communication, not information per se.

Many have run with his theory, applying it to information in general as though it doesn't matter whether there are senders or receivers since we can quantify bits with or without interpreters of those bits. Bits are defined broadly enough to measure potential signs. Potential signs are not inherently informational. To understand how potential signs get interpreted, we have needed an explanation for selves that aim to interpret.

Charles Sanders Peirce founded the field of *semiotics*, the theory of signs or significance that yielded the distinction between icons, indexes, and symbols. Recognizing the problem with simply assuming there are selves, he developed a theory that doesn't include them. For Peirce, information was a three-way relationship between a sign, object, and interpretant. An interpretant is not an interpreting self or person, nor is it a self's end-directed aims. Rather it's whatever physical change results from the interaction with the sign, for example, braking in response to a stop sign.

Cause-and-effect events link two events, thus creating series or sequences of one-to-one links, a cause-and-effect chain of events. In contrast, sign relationships, by Peirce's account, link three events: object, sign, and interpretant. Its three-event structure makes branching possible, sequences of signs about objects producing interpretants that then become signs or objects for other interpretants.

Like many of his followers today, Peirce worked to identify the range of possible object-sign-interpretant relationships, starting with his distinction between icons, indexes, and symbols. Peirce's reliance on interpretants as physical changes enabled him to sidestep the mystery of purpose.

At times in his career and especially toward the end, Peirce intimated that sign relationships might be evident in all phenomena. By not distinguishing selves from nonselves, he left ambiguous which domains have sign relationships. At some points his work seems to support eliminativism (no selves anywhere) and, at other points, panpsychism (selves everywhere throughout all physical phenomena).

Another researcher, Gregory Bateson, contributed an insight to our understanding of information when he described it as "a difference that makes a difference."[3] Read literally and taken out of context, this definition doesn't distinguish information from energy, since work is also a physical difference that makes a difference, for example, a difference in a cue ball's movement making a difference to an eight ball's movement.

But Bateson meant the second difference as a double entendre: both a different interpretant in Peirce's terms (a different behavioral response) and a difference of value, a good or bad difference with respect to aims. By implying aims, Bateson implies selves, but doesn't specify how selves actually use information to "make a difference." Without a solution to the mystery of purpose, "a difference that makes a difference" can imply that naked RNA theory yields information, as though some molecules replicating faster than others is information since it makes a difference to replication."

HOW INTERPRETATION HAPPENS

With Deacon's hint at how to solve the mystery of purpose, we can begin to build out from these worthy efforts and others to a credible science of information as signs interpreted by selves, given their aims.[4]

Extrapolating from Deacon's approach, presented in articles,[5] I would argue that interpretation depends on three levels of constraint in a self's interaction with its environment: constraints implicit in initial expectations, constraints on interpretations of divergences from expectations, and constraints on responses to divergence from expectations. These constraints may be evolved or learned, depending on the kinds of selves involved. Either way, they enable a self to correct for divergences that would otherwise thwart its aims.

To take a simple example, you turn on the hot bath water and put your hand under the faucet. You wait, expecting hot water to flow. If it doesn't, that's a sign to you about something, even if you don't know what. But you have guesses about the range of likely explanations.

The divergence from expectations, cold water flowing, points to something intervening in the expected sequence of events. At first, you don't know what the divergence is about, but you have constrained expectations regarding what might have intervened, for example, the likelihood that it's a blown fuse, the old water heater on the blink, an unpaid energy bill, or someone else having used up all the hot water with a long bath. By trial and error, you narrow in on what eventually gets you your hot bath.

First, there's the expectation that has you testing the water for temperature. You don't assume it's impossible that the water from the hot water valve would stay cold, just unlikely. You intuit a probability distribution, the constrained relative likeliness of some outcomes compared to others, given your accumulated habits of expectation based on experience.

Second, when hot water doesn't flow, you have a constrained range of expected explanations of what has intervened, in other words, what the absence of hot water points to, or is about. Third, you constrain these possible explanations by trial-and-error responses, until you restore your ability to get your hot bath.

DIVERGENCE FROM EXPECTATION

Shannon's fundamental bit unit is fifty-fifty, "either this or that," in other words, one of two equally likely possibilities. By this definition, a one-of-one possibility is zero bits, no communication, information, or news. If you get a communication that you're 100 percent certain you will get, it's not information. By this definition, bits measure surprise or news, differences that make a difference to selves by diverging from expectation.

If there's a 1 percent chance that the water won't get warmer, and it doesn't, then that's information, a significant divergence from expectation. Divergence is what captures our attention, especially when it threatens to thwart our aims. Obviously, a hot bath is not a self-regeneration requirement, but for humans aims can extend well beyond self-regeneration.

With the bathwater, the first constraint in the interpretive sequence is a Shannonian constraint applied not just to a potential sign but to an interpreted one: Of all the expected states the water flow could be in, it's in a narrowed range of states, staying surprisingly cold. It's not a pinpointed state but a range, anywhere below a threshold of expected warmth.

Shannon's model for communication involves a signal running through a channel, which is actually a physical medium that can be altered by things outside it. Although the bathwater is flowing through a pipe, this is not the information channel. The water is the medium in question, and the entire system that reliably provides hot water most of the time, including the water heater, the gas lines or electric wires, the public utility, the other users of the water, and the like, provides the possible sources of system failure that the water temperature could be about for you.

When the water temperature diverges from the expected range, you interpret it as a sign that something in this largely reliable network of influences has been altered. This is the indexicality discussed earlier, a pointer to some about-ness, something outside the sign medium itself. And you have a constrained range of guesses about what it might be.

Through a trial-and-error process of elimination, you *hone in* on a pragmatic response to the unexpectedly cold bathwater. The process ends with your restoration of hot water, an interpretation that enables you to engage in a course of action to recover from the perturbation that thwarted your prospects of achieving your aims.

Calling this constraint-based approach to the interpretation process *honing in* is a bit more accurate than the conventional term *homing in. Home* originated with homing pigeons that targeted a pinpointed home when far away, and was later apply to missiles that home in on targets. Missile homing became a central focus in cybernetic research, which focuses on how, through negative feedback loops, a mechanical device homes in on a predesignated target. For example, like targeted missiles, thermostats home in on a predesignated set point, a temperature, an aim toward a target state specified by its user or designer. Through cybernetics, it was possible to equivocally assume or assume away selves, implying either that thermostats are no different from selves in that they aim or that selves are mere cause-and-effect mechanisms like thermostats.

Given Deacon's constraint-based approach, *honing* is a better choice than *homing*, since it implies a narrowing process of elimination and does not assume a predesignated, pinpointed target.

In the broad generalization of the interpretive process described now, we see three nested constraints, each a kind of honing, not by targeting a pinpointed bull's eye, but by constraining possibilities through processes of elimination. First, there are constrained expectations for what is likely to occur; second, constraints on likely sources of the surprise divergence from expectation; and third, the process of elimination by which you arrive at a response that restores expected results.

Following the hot-water incident you are likely to update all three kinds of constraints, adapting habits of expectation to better fit your circumstances. For example, if you discover that the hot water was absent because your new housemate used it all up with a long bath, you may come to expect that particular cause of cold water when you go for future baths.

FORESIGHT AND EXPECTATION

Only selves have expectations, their accumulation of evolved and, in some organisms, learned habits of interaction given their aims. Expectations are implicit in a self's fitted, functional traits. A self's fitness to an environment is a prediction, an anticipation of what will work to maintain the self. Most expectations are far from conscious and deliberate, yet they are literally *deliberated.* That is, they are reductions in possibilities in ways that are representative of a self's environment and what the self expects will work in it.

Self-regeneration is the first aim and the most fundamental of such anticipations or expectations. Even the minimal autogen has this most basic kind of expectation, even though it lacks the capacity to respond adaptively when expectations aren't met.

Given its protective capsid and ability to self-repair when broken, it responds to "expected" perturbations (that is, those that are probable and have recurred repeatedly over the course of an autogen lineage). Its self-regeneration is suited for environments in which reactants are intermittently available and where breakage is possible.

To put the minimal autogen's habit into words, which of course it cannot, its self-regenerative capacity to anticipate translates to "If broken, then repair and reproduce." This is a habit that works and is likely to continue working, in that it maintains its self-regenerative constraints. Its constraints are functional traits or habits. Its habits are both means and ends—habits likely (expected) to serve to maintain its habits into the future.

The selective autogen adds another expectation, a capacity to respond to different circumstances. Again, to put it into words, if sufficient numbers of reactants are present, then open; if enough reactants aren't present, stay closed.

Both of these expectations are predictions about what is likely to occur, loosely parallel to your expectations that hot water will flow when you open the hot water faucet.

In the selective autogen, we find an evolved capacity that loosely parallels your learned response when hot water doesn't flow. For the selective autogen, reactants binding to its capsid surface point to the likelihood of reactants being present. This is an evolved habit of expectation. An expectation is an anticipation of what is likely to occur in the future.

Note that self-repair across all selves has the features we find in the bathwater example. A symptom is a divergence from expectation. The presence of a symptom doesn't point precisely to what is causing the symptom. By trial and error, a self explores for possible remedies to alleviate the symptom. Remedies may stop short of addressing the true source of the symptom, as when palliatives alleviate the symptom without addressing its cause. There are parallels to be drawn from the minimal autogen's capacity for self-repair all the way out to a human coping strategy for remedying anxiety.

GENERALIZING ABOUT INTERPRETATION

Obviously, the bathwater example and loose parallels to the autogen as origin of interpreting selves do not encompass all features of interpretation. This was merely an example illustrating a direction Deacon takes in articles and in a forthcoming book that will provide a far more detailed and general theory of the interpreting self. Still, from this brief sketch there are a few key points worth noting:[6]

Interpretation makes a difference to selves, given their aims. It is motivated, not merely a process of registering data. It is driven by habits of expectation, and anticipation, broadly defined to include everything from evolved adaptive traits in any self to updated intuitions in human selves.

Interpretation is a process of elimination played out on three levels of expectation: constrained initial expectations, constrained expectations regarding divergences from expectations, and constrained adaptive expectations about responses to such divergences.

A sign does not contain its about-ness. Its about-ness is honed by a self's interpretation process, most fundamentally to overcome perturbations that would impede achieving its aims.

A sign doesn't cause effects in the simple and direct manner that physical causes produce reliably predictable effects. With causes and effects, both are present and directly interacting. With interpretation, the sign can be the absence of something expected—in our example, the absence of expected hot water.

The self's responses to signs are the products of pragmatic guesswork (evolved or learned) about how to recover from perturbations, guesswork that halts when recovery is achieved and not necessarily when the true source of the perturbation is discovered.

Interpretations are, for this reason, inherently fallible. The true source of the perturbation may be different from the identified source. For example, for millennia people thought disease and disasters were perturbations caused by angry gods, avoidable through sacrifice and appeasement.

VII

IMPLICATIONS

27

A CONSTRAINT-BASED APPROACH TO EVOLUTIONARY THEORY

REDUNDANCY AS FREEDOM TO EXPLORE

In this final section, I'll be exploring possible implications of Deacon's theories, some of which are suggested by Deacon, others only hinted at, and still others my own extrapolations starting first with possible implications for evolutionary theory.

Researchers are right to point out the crucial role played by template molecules. As you'll recall from our discussion of Shannon's information theory, redundant signals tolerate greater noise without loss of signal. Having a copy of a signal for safekeeping enables us to experiment without losing our place. For example, if you are experimenting with a radical rewrite of a text you're working on, having a redundant backup copy gives you something to go back to should the experiments fail. Genes are redundant representations of a body's dynamics. Both represent what works for a self within the environment.

In the template autogen, the protection function in self-regeneration becomes partially offloaded onto the template, allowing for greater evolved variability in the autogen's dynamics. We see this from the start in the way that the template autogen's dynamics are freed to explore a wider range of autocatalytic set members and sequences than are possible without a template. Redundancy relaxes the constraints on the dynamics, tolerating greater variance and making it possible to accumulate a wider range of adaptations that would then be recorded in the template.

A conventional *law-of-effect* approach to evolutionary theory attributes heritable variation to chance genetic mutations anywhere within the genome. The mutations do not aim for fitness. Rather, say biologists, fitness is a product of *selective pressures*, the demands a selective environment imposes that result in greater survival of those lineages that supply whatever meets natural selection's demands. The term is problematic, since *selection* implies selves choosing to suit their aims, and *pressure* is a term from physics. The literal impression is that natural selection wants traits and then imposes physical pressure on organisms or genes that pushes them into compliance with selective demands. Still, the core concept is sound. Selective pressure is actually aimless environmental constraints that prevent a range of selves from sustaining self-regeneration.

Such environmental constraints vary in degree, from tighter to looser tolerance in the engineering sense. Tighter tolerance constrains survivable selves or traits to a narrower range of variability. Looser tolerance constrains to a wider range and increased variability. Of course, the word *tolerance* also implies aims. It's not as though natural selection is capable of tolerating or not tolerating in the way a self would.

Nonetheless, the point here is that we should think of natural selection not as a promoter or demander of traits, but as a constraint on selves, which promote their own persistence through self-regeneration. For simplicity, we'll stick with the term *selective pressure*. Selective pressure varies in its strength—higher pressure indicating less tolerant environmental constraints.

To explain adaptive traits using the most common approach to evolutionary theory, one focuses on the onset of strong selective pressures, which, like a self, favor some lineages over others. Taking this approach to its logical conclusion, one can simply identify a trait, identify the selective pressures that would make that trait advantageous, and simply assume chance mutations produced the trait in response.[1]

This approach satisfies our preference for simple explanations. It can make evolution seem like cause and effect of the simplest kind. Natural selection makes demands that cause the survival of chance genetic mutations that can supply to the demand. We hear this kind of simplification in the arguments that evolution designs bodies or programs DNA, or in evolutionary psychology, in which often what counts as a full explanation for a behavior is that there was a possible advantage to it, a natural selection demand that was supplied by random DNA mutations, an approach that paleontologist Stephen Jay Gould scorned as "just-so stories."

Before Darwin, the primary speculation about how traits evolved was Jean-Baptiste Lamarck's suggestion that traits evolved through use. Giraffes developed long necks by stretching to reach higher branches. The exertions of parents were somehow thought to pass on to children, as though building muscles in your lifetime would mean your children would be born with bigger muscles. Lamarck's theory was discredited as an explanation for biological evolution, though there may be some application to human learning.

Through just-so stories, however, Darwin's theory can be interpreted just as simplistically as Lamarck's. Faith in the law of effect makes people confident that necessity is the mother of adaptation. It's a misunderstanding of evolution to identify what an environment might have "demanded" and to simply assume that random mutation can supply it.

EVOLUTIONARY DECONSTRAINT

Natural selection doesn't impose demands. Natural selection is an aimless, passive process of relative elimination of the unfit. Only selves engage in processes of production, a point often overlooked in many conventional interpretations of evolutionary theory, though not in Darwin's emphasis on organisms' "struggle for existence," selves aiming to self-regenerate, though, in most organisms, not aiming to tailor or adapt for improved fitness, which is a quality most pronounced in the human aim to adapt through learning.

Self-regeneration, the self's struggle for existence, is not the only point sometimes overlooked in interpretations of evolutionary theory. Another is that mutations alter genes that already yield functional traits. In so doing, they yield variations on existing traits.

Also, though mutation is constant enough across time that we can use the rate of mutation as a "genetic clock" in cladistics (research into genetic relatedness over time), mutations are more likely to accumulate when selective pressure is low than when it is high. In other words, when environmental constraints or tolerances are looser, mutations can accumulate within a lineage without risk to self-regeneration.

Furthermore, in just-so story, demands-cause-supplies explanations for adaptations, one tends to pay attention only to the onset of selective pressures, and not to when high selective pressures are reduced, meaning when tight tolerances are loosened. Sometimes a trait necessary for survival stops being necessary; as a result, mutations can accumulate within already functional genes without risk to survival. These accumulating mutations sometimes generate variations on once-functional traits that then may become repurposed for other functions.

Selective pressure can become reduced or relaxed by a variety of means. For example, selves might migrate to new environments where resources are more plentiful or predators are scarcer. Or new environments may provide externally what selves would formerly have had to produce for themselves.

As is well understood in evolutionary biology, pressures can also be lifted by production of multiple copies of the same gene, for example, through gene duplication or an increase in the number of times that a gene is expressed to produce its encoded protein. With more than one copy of a gene, there is less selective pressure on each copy to continue to fulfill its former role.

We should keep in mind that while we tend to focus on qualitative changes resulting from genetic mutation, there are quantitative changes also, redundant production of the same gene product (usually a protein), reducing the burden on each copy of the gene to continue to maintain its original function.

We can describe this sequence of events as *redundancy*, *relaxation*, and *possible repurposing*.[2] I'll call them the three Rs.

THE THREE Rs

Redundancy means having multiple ways to achieve an aim. Redundancy leads to *relaxed* selective pressure, making it possible to maintain self-regeneration despite accumulating mutations. And relaxation may lead to possible *repurposing*, the accumulation of variations in already-functional genes that can give rise to new functionality.

For an intuitive parallel, consider what happens when an organization overstaffs to complete some task (redundancy). With more staff than necessary, each staff member is under relaxed pressure—not relaxed genetic pressure of course, but relaxed social and economic pressures.

Under relaxed pressure, some employees might drift toward nonproductivity. But others might diversify, repurposing their already-honed functional skills to related tasks. And there are often related tasks. A lot of diversification occurs through variation on a functional theme.

Sometimes, the redundant employees' repurposed functionality proves advantageous to the organization. For example, a staff person, under relaxed selective pressure, might stumble on an innovation that becomes an essential source of productivity and efficiency for the organization, thereby adding to the organization's capabilities. Through this innovation, new synergistic couplings can arise within an organization.

Many organizations institutionalize redundancy in order to cultivate innovation. Businesses have R & D departments and offsites, and universities offer sabbaticals and long vacations, all under the assumption that, with relaxed pressure, functional staff may produce desirable innovations.

But back to biology: attention to the three Rs offers potential insights for how a body's intricate divisions of labor evolve over time.[3]

AN EVOLVED FRANKENCELL

When engineers design functional machines, they bring together functional parts from disparate sources. As such, engineering is much like Dr. Frankenstein creating his monster by putting together functional parts gathered from various bodies.

Biological evolution rarely innovates the way that engineering does. In biology, parts generally diversify from within; they're not brought together from without. Redundancy, relaxation, and repurposing provide insights into the biological alternative to the engineering or Frankenstein metaphor, parts brought together to achieve functionality.

However, for a rare example of a Frankenstein-like biological process, consider the *endosymbiont hypothesis* for the evolution of eukaryotic cells, a process that also is better understood through appreciation of the three Rs. According to the hypothesis, one formerly independent cell engulfed others. One engulfed cell, a protobacteria, became the engulfing cell's mitochondria; another, a cyanobacteria, became the chloroplasts in plant, an algal eukaryotic cells. The conventional evolutionary interpretation focuses on the adaptive demands met by this innovation.

But if there were not an immediate adaptive advantage to combining, what would keep the engulfed and engulfing cells together over generations?

Deacon argues that it would be relaxed selective pressure. The two cells would have redundant capabilities, which would therefore be under relaxed selection, looser tolerances that allowed mutations to accumulate in each, degrading each cell's ability to survive independently. We see the effects of relaxed selection in the engulfed cell—the mitochondria's loss of autonomy and functional constraints. Though the engulfing is not a function of relaxed selection, its consequences were just each formerly independent cell losing functionality and therefore ending up in symbiotic or codependent relationship with the other.

With accumulating mutations, each cell's already-functional traits could become repurposed toward specialized functionality, a division of labor, yielding the internal codependency, the synergistic coupling we find in eukaryotes today. In eukaryotes, the cells and their mitochondria are unable to survive without each other. Still, through their synergistic coupling they have gained many adaptive advantages.

The three Rs are evident in the evolved specialization of duplicated body parts—for example, the specialization within a series of limbs, digits, bones, or teeth. With multiple copies of the same part, each part is under less selective pressure to maintain its original function. Under reduced selective pressure, a copy can accumulate mutations and in the process repurpose toward new synergistic specializations. For example, a lobster's series of opposable limbs become specialized to the divergent roles of antennae, claws, forelimbs, hind limbs, and spinnerets, which females use to carry packets of sperm awaiting egg-laying season.

Chance genetic mutations are still essential to the explanation for such specializations, but the explanation is incomplete without an appreciation of the role played by the three Rs. If it weren't for the multiple copies of parts and the resulting reduction in selective pressure, mutations affecting already-functional traits would have had a lower chance of yielding new specializations. Redundancy allows mutations to take a random walk that lands them in new, specialized functionality.

As long as self-regeneration is maintained, there exists the potential for redundancy, relaxation, and repurposing, which in turn leads to added synergistic couplings, greater differentiation and division of labor within bodies, and the accumulation of synergistic couplings from within.

DIVISION OF LABOR BETWEEN SELVES

Selves depend upon one another in many ways, given a variety of names depending upon the kind of adaptive advantage seen as driving the dependency. Here are a few: mutualism, symbiosis, parasitism, predator/prey, extended phenotype, eusociality, sociality, pair bonding, kin selection, reciprocal altruism, group selection, care, and even love. When these terms are evoked or debated, the process of distinguishing among them often focuses simplistically on identifying the kind of natural selection demand that "causes" the supply of the trait, given random genetic mutations.

In understanding the divisions of labor between selves, we might benefit from paying more attention to the role played by the three Rs and the resulting synergistic coupling.[4] Here we will not address the details of these variations on the theme of between-selves divisions of labor, or debate the merits of describing any as mutualism or another term. Rather, we'll visit one relatively simple example, increased codependency between organisms.

Ascorbic acid is an antioxidant that protects bodies from the corrosive effects of oxygen. Like almost all mammals, our primate ancestors produced their own ascorbic acid. Anthropoids—monkeys, apes, and humans—no longer do. We now depend upon external sources of ascorbic acid.

About sixty million years ago, primates found their way into trees, a new environment that relaxed selective pressure for nocturnal foraging. In trees, they could forage by day, living on insects and small lizards without threat from predators. Trees also bore fruit in a symbiotic relationship with birds that disseminated seeds, a functional benefit to both.

About thirty-five million years ago, the anthropoid diet became increasingly focused upon fruit. With two sources of ascorbic acid—one internal, the other external in the reliable supply of fruit—anthropoids were under relaxed selective pressure to maintain their internal production of ascorbic acid.

Anthropoids today still have the genes that in other mammals facilitate the production of ascorbic acid, but in us, these genes have become junk DNA, no longer functional, having mutated without consequence to survival. Since the genes weren't repurposed, this is an example of redundancy and relaxed selection that did not result in repurposing.

Having lost the capacity to synthesize their own ascorbic acid, the anthropoids thus fell into a symbiotic relationship with fruit-bearing trees.

Their dependency on fruit imposed new selective pressures that led to the evolution of the anthropoids' ability to forage on the outer limbs of trees, find the sugar-rich and slightly acidic content of fruit attractive, metabolize the sugars in fruit, and tolerate the ethanol that ripe fruits contain. In this manner anthropoids and fruit trees entered into a synergistic coupling, a between-selves codependency mutually beneficial, mutually constraining, and also mutually freeing. Anthropoids no longer needed to maintain their ability to autosynthesize ascorbic acid; fruit trees were no longer wholly dependent upon birds for broad seed dissemination.

And this symbiotic coupling led to other symbiotic couplings, again by the three Rs. Prior to arboreal life, primates foraged at night, depending upon eyesight that captured as much light as possible. Daytime foraging relaxed selection for night vision. Three-color reception resulted from the three Rs since the gene producing rhodopsin, the protein that enables light reception, located on the X chromosome duplicated to three copies. With this redundancy and therefore reduced selective pressure, each of the three lost rhodopsin's original broad spectral tuning maximized light capture for night vision. Each acquired mutations that changed its optimal sensitivity to a different narrow range. The result was a synergistic-coupling division of labor, which made color vision possible, thought to originate as a way to differentiate ripe from unripe fruit.

This story illustrates the range of variations possible in evolution. Arboreal foraging was a new environment that relaxed the selective pressure for night vision and other traits necessary for nocturnal foraging and predator avoidance. For fruit trees, anthropoids became redundant to birds as means for seed dissemination.

The ascorbic acid in fruit relaxed selection on the anthropoid's internal production of this essential nutrient, in this case not toward repurposed functionality but toward extinction of the capacity to autosynthesize ascorbic acid, which, in effect, trapped the anthropoids into dependency upon external sources of ascorbic acid. Dependency on fruit imposed new selective pressures that resulted in anthropoid color vision among other functional capacities.

It may seem surprising that human color vision probably originated in our lineage as a means to distinguish ripe from unripe fruit. After all, we now rely on color vision to serve a wide variety of aims, of which fruit differentiation seems relatively trivial. Once selves have a trait, it can be repurposed sometimes without sacrifice to the original aim.

An approach to Darwinism that takes into account not just the onset but also the relaxation of selective pressures provides a richer, subtler, and more fruitful approach to evolutionary explanation. It moves us away from the confusions that arise from thinking of natural selection as a programmer, designer, cultivator, or demander, in other words, a self with aims. At the same time, it moves us away from treating evolution as paralleling cause and effect in simple deterministic relationships, as when we say that selective pressure causes the production of a trait.

IMPLICATIONS OF RELAXATION FOR ALTRUISM?

The three Rs may also contribute to an overlooked explanation for altruism, developed by Deacon with colleague Julie Hui.[5] When selves lose the capacity to produce for themselves what is reliably provided by other selves, they become dependent on others much as we became dependent upon fruit trees for ascorbic acid.

Once selves become dependent, new selective pressures expose them to new threats and opportunities such as those that led to the evolution of anthropoid color vision. The root source of these new selective pressures was relaxed selective pressure to autosynthesize ascorbic acid, which trapped us in codependency.

Dependency doesn't necessarily start with adaptive advantage but may begin with relaxed pressure to produce what is available externally. Many kinds of mutualism, including sociality, may sometimes begin with relaxed selection, the ties that bind us originating in lost autonomy rather than the onset of selective pressure.

Today, many of us live in cultures that celebrate individuality, and many of us feel as though we have achieved a high degree of autonomy. Still, this impression of independence is a function not of true independence, but of the unprecedentedly high reliability of goods and services supplied to us by others upon whom we depend. Reliability enables us to ignore our dependencies. For example, one can ignore how we get our hot water for baths, since it appears so reliably.

Our modern, highly reliable technologies do not make us less codependent, but rather less aware of how codependent we are. As in our bathwater example, we don't notice when the expected happens. The better our technology gets, the more the expected happens. We don't attend to all sorts of dependencies on others when others reliably and efficiently provide what we need.

Perhaps, then, some kinds of altruism result from such mutualism, originating in redundancy, relaxation, and repurposed specializations that together result in divisions of labor between people and therefore growing mutual dependencies, a net-work of social synergistic couplings.

SHIFTING CONSTRAINT AND DECONSTRAINT

The role of redundancy, relaxation, and repurposing made possible through gene duplication is well understood in biology. Deacon argues that the three Rs have application beyond gene duplication, as in the anthropoids' redundant sources of ascorbic acid.

There are also social implications, played out not in heritable genetic traits but in social, cultural, and personal aims, commitments and cares and how they shift. Over your lifetime, aims that once were important to you no longer are, while other aims of no importance to you have grown to become priorities. The three Rs may help explain some of these.

For a social-historical parallel to the ascorbic acid story, consider agriculture. Originating about nine thousand years ago, farming provided efficiency in food production, making some hunter-gatherer producers redundant. Under relaxed socioeconomic pressure, some food producers repurposed their skills to other activities upon which early civilizations came to depend in an ever-growing, ever-shifting division of labor and, as a result, social synergistic couplings.

More recently, large-scale farming, mechanization, and agribusiness have made our individual capacity to grow our own food redundant. Since food markets became reliably available, farming was no longer a do-or-die skill, and now many of us couldn't garden if our lives depended on it. Farming atrophied in response to relaxed pressure, not at the genetic level but in the loss of culturally inculcated skills.

When modern humans were freed from farming, new pressures and opportunities emerged. For example, we have become dependent on income for buying groceries and vehicles for transporting food to and from grocery stores.

This is a commentary not on modern life but on how things change, not just by the reliable laws of cause and effect, and not just genetically, but in the unreliable vagaries of changing aims within our established, growing, and changing networks of dependence. Since selves are crucial features of one another's environments, when some selves' aims change, they change circumstances for other selves. What was once significant can become insignificant and vice versa.

Selves fall in and out of dependencies over the course of evolutionary history, but also, and especially for humans, over the course of cultural and personal history. The shifting of aims shifts into high gear with humans, given our capacity to interpret symbols, which affords us a way to anticipate, speculate, and venture toward futures and technologies shaped by our aims in ways unprecedented in other species.

28

IMPLICATIONS FOR THE FREE WILL DEBATE?

NEITHER FREE NOR DETERMINED

The free will versus determinism debate is often framed as though the only source of events were cause and effect. Framed this way, the question becomes whether a self can ever act as an unmoved mover, the initiator or cause of actions that does not have prior causes. If we can demonstrate that the self acts as an uncaused cause, we have free will. If we can't, then there is no free will.

To illustrate, an oft-cited study that some argue suggests the absence of free will demonstrates that neural activity precedes conscious decisions to act, suggesting that the cause of human action is not a conscious choice but material neural dynamics.[1] By many accounts this study suggests that we are machines, not ghosts with wills.

Deacon's approach suggests unexplored aspects of what we intuit as free will. First, he suggests that we should seek the roots of will in any self's aims, not in humans, where most of the free will debate is focused. Nor should we just seek the roots of will in quantum mechanics, where the departure from mechanistic determinism is sometimes regarded as evidence for a kind of free will.

Deacon's proposed solution to the mystery of purpose suggests how will enters the universe as aims, inherent in all selves.[2] From the autogen forward there are dynamics working against abiotic dynamics. We channel energy into our effort to maintain self-regeneration despite the second law tendency toward irregularity. Deacon argues that what is new with selves is a capacity for "novel work" in the universe, the work that furthers the selves' aims.

Second, our brief exploration of interpretation suggests how aims depart from the predictability of cause-and-effect events in classical physics, because interpreted signs don't work the way material cause-and-effect events do. Again, a stop sign doesn't cause you to stop unless you crash into it. A stop sign is open to interpretation, and this openness, writ large, is a fundamental source of unpredictability in life, especially in humans.

In our brief exploration of the origins and nature of interpretation in iconic signs, we saw how indifference is the root of interpretation. The minimal autogen emerges indifferent to all differences. It has no capacity to interpret any distinctions within its environment. The selective autogen evolves a modest capacity to distinguish environments with and without available reactants.

It does so by means of a threshold effect. If enough reactants bind to its surface, the capsid is likely to break open. Of course, other selves have more refined capacities to distinguish features within their environment but the distinctions are never perfect. There will be false positives and negatives, errors, potentially life ending. Interpretation is always imperfect and the survivors are those that interpret well enough.

FROM THRESHOLDS TO CATEGORIES

It's easy for us to think that selves are programmed to target pinpointed states, for example, that a bacterium has internally programmed if-then statements, such as "if sugar present, then move."

A bacterium doesn't function based on categorical distinctions like those implied by terms like *sugar* and *move*. Bacteria aren't programmed in the functionalist sense that requires symbols, language that enable humans to distinguish categories.

Among selves, humans have unprecedented symbolic competency. Our socially conventionalized systems of signs make it possible for us to label what's on either side of a threshold. With language, we can categorize subtler distinctions between things on one or the other side of a threshold.

We have categorical names that fill dictionaries, terms for every relevant distinction in our environments and our imaginations, and when we don't have terms for something within our symbol systems, we can add them. Symbol systems, and in particular languages, give us a capacity to mentally model any aspect of our environment, experiences, and imaginations, not just with threshold-based distinctions but also with named categories for things, events, concepts, and experiences.

It is this symbolic capacity that makes us unprecedented in our ability to model the world within our symbolic systems, systems that take on a life of their own within us since all of the terms are networked to one another. We have, in effect, a thesaurus in our minds, making it possible for us to imagine what isn't immediately present because it's represented already in our internalized networked systems of symbols.

UNEXPECTED CONSEQUENCES

With interpretation, and especially symbolic interpretation, comes a consequence that one might mistake for free will. There are aims and interpretations that shift and, in turn, shift other aims and interpretations.

With language, indifference to differences remains an issue, indeed an expanded one. As humans go around categorizing things, we overlook differences that may prove significant. Our categorical names for things outline some distinctions but not all. We may say of two options, "These are equivalently *worthy aims*," dedicate ourselves to one, and end up in very different circumstances than had we dedicated ourselves to the other.

We commit to dependencies based on a categorical sense of what we're getting into, but these commitments result in unexpected consequences for better or worse. Either way, our aims and interpretations shift. This is so for all selves. It's a large part of what makes evolution so unpredictable.

But with our symbolic capacity to categorize and generalize, rooted in attention to some differences while ignoring others, unpredictability expands exponentially. Seemingly insignificant differences may turn out to make significant differences, and seemingly significant differences may turn out to be insignificant. Generalizing to categorical distinctions, we don't know all of what we're getting into. Having gotten into something, our aims shift in unpredicted ways.

Suppose you take one of two jobs, both of which fit the general category of value for serving your aims, but each with different consequences, some unexpected given your generalizations. The different consequences impose different demands upon you, different incentives and disincentives, which are the learning counterpart to selective pressures.

With life and especially human life, novel work is generated. We evolve and we learn in unpredictable ways that can appear like free will, very different from the reliable cause-and-effect predictability we find in classical physics.

A GROUNDING IN INCOMPLETE DETERMINISM

Heisenberg's uncertainty principle demonstrates at the quantum scale that it is impossible to measure cause-and-effect events accurately, because to measure a particle's position compromises our measurement of its momentum, and vice versa. To some, this suggests that there may be determinism at the quantum scale that we can't resolve simply given our limited capacity to measure what occurs.

Deacon and others, including Nobel Prize–winning physicist Ilya Prigogine,[3] reject this argument. Instead, they argue that it is time to abandon the determinist assumption that has guided science for centuries.

At all scales, infinitesimal differences can amplify toward larger differences, so whenever we're tracking a sequence of cause-and-effect behaviors, we cannot predict the ultimate outcome with absolute certainty.

The source of amplifications, whereby little differences can make big differences, is the solution to what Prigogine calls the time paradox: "How can the arrow of time emerge from what physics describes as a time-symmetrical world?"[4] In a strictly deterministic universe, one should be able to play the movie of events forward and backward, yielding time symmetry, but all evidence suggests that time is not reversible in this manner. This is most obvious with selves and aims.

Prigogine's solution has much in common with the concept of self-organization or emergent regularization as described earlier. Here we will ignore the details of his approach and concentrate instead on his conclusions. Prigogine argues:

> In solving the time paradox, we also solve the quantum paradox, and obtain a new, realistic formulation of quantum theory. This does not mean a return

> to classical deterministic orthodoxy; on the contrary, we go beyond the certitudes associated with the traditional laws of quantum theory and emphasize the fundamental role of probabilities. In both classical and quantum physics, the basic laws now express possibilities.[5]

We don't have determinism. We have incomplete determinism, whereby differences can accumulate toward a degree of inherent unpredictability not just in the eyes of the observer but also in reality itself.

Materialism and determinism tend to be assumed together. A billiard-ball cause-and-effect universe would be a deterministic universe, a universe in which Laplace's demon with infinite calculating power and knowledge of all material objects could compute all past and future events time-reversibly.

This is not our universe. Increasingly, physicists recognize our universe as probabilistic from the ground up—a fundamentally nondeterministic universe. Determinism is an idealization, plausible in mathematical and computer modeling, but not in nature.

Grounding physics in quantum mechanics, we are compelled to reject determinism at all scales of analysis. Philosopher Ernest Nagel summarized the point, arguing, "It is impossible to trap modern physics into predicting anything with perfect determinism because it deals with probabilities from the outset."[6]

How, in such a universe of unpredictable possibilities, could some dynamics become more probable than others? By emergent constraint, the amplification of regularities, and new constraints that arise from throughout dynamic interaction.

MOST BUTTERFLIES DON'T

Amplifications within a universe of incomplete determinism may bring to mind the "butterfly effect," the argument that a butterfly flapping its wings in the Amazon could, with the right initial conditions, amplify into winds that alter weather patterns far away. People have usually taken this as arguing either for total uncertainty or for hope that the little differences they aim to produce will have great impact. A more useful takeaway is that initial conditions matter.

The vast majority of butterfly wing-flapping generates an infinitesimal flutter in surrounding air that quickly peters out as irregularity due to the second law, yielding no weather-altering change. It would take very rare initial conditions for the flutter to amount to anything more. Amounting to more would result from emergent regularization whereby a small fluttering compounds to a big fluctuation. Close to those initial conditions there's what scientists call *sensitivity to initial conditions*, the slightest difference in initial conditions making a big difference as to whether the fluctuation amounts to anything at all.

Initial conditions for the origin of self-regeneration would be rarer still. Life anywhere in the universe probably starts very rarely. Still, when it starts, the probability of it continuing is increased. The longer life continues, the greater the probability that it will persist. Life here on Earth is much more robust than it was at its origin. It's very unlikely that a butterfly's flapping will change the world climate. But life in general has changed lots, including world climate.

There's something likelihood-increasing about selves and their aims that we must therefore explain. How from some rare difference in initial conditions could selves grow to make such a big difference? Most, but not all change in the universe is ephemeral, but that very much changes with life.

At any scale, little differences in initial conditions can amplify into big differences, as the butterfly effect suggests. Still, at the quantum and even the classical scale, the vast majority of minor variations don't end up making big differences, and when they do, very rarely does a subtle difference in initial conditions make one of several possible big differences.

Set aside butterflies, since they too are selves. A minor wind fluctuation resulting from a wave crashing might emergently regularize into a hurricane. If it doesn't, it's likely to peter out rapidly. It's very unlikely to amplify into one of several alternative hurricanes.

Within living bodies, synergistic couplings honed to serve aims over eons of evolution are far more likely to amplify in alternative big differences, and in us, symbolically competent humans, all the more so. There may be more wind generated by butterfly wing-flapping than by a world leader's yes or no head movement on some pivotal question, but what amplifies out from the leader's movement is more likely to be amplified in different directions than the butterfly flapping.

THE LOGIC THAT MAKES LOGIC NONDETERMINISTIC

Paradoxically, our symbolic capacity breeds confidence in our ability to predict change, even while reducing predictability. For example, we formalize mathematics, a symbolic system giving us the impression of a clockwork universe. Formulas depend upon categorized variables, which must be defined and filled by means of interpretation. Looking at the bare syntactic structure of formalizations, we can get the impression that we are modeling a inviolably determinate reality, but only if we ignore the role interpretation plays in applying formulations to reality.

The same is true for logic, a formalized system that, stripped of application, appears to be rigidly determinate, but only if we ignore the role that interpretation plays in replacing variables with identities.

Logical deduction seems determinate. If all men are mortal and Socrates is a man, then we know with certainty that Socrates is a mortal. But how do we know that all men are mortal? By induction, generalizing, and categorizing from instances, since all men that we know of from the past have proven to be mortal.

But not all men have proven mortal yet in that there are many still alive today. The past is often prologue, so we build habits of interpretation from it, as Pavlov's dog did with the bell. But consider the chicken that assumes by induction that farmers bring food, until one day the farmer brings an ax instead. Induction is inherently incomplete, open to interpretation.

And how do we know that Socrates is a man? By what philosopher Charles Sanders Peirce called *abduction*, finding traits in common between instances of men and Socrates. In other words, iconism: this and that appearing to be in the same category because they have traits in common. But how many traits in common must we witness before we can interpret confidently? It's possible to interpret that someone is a man and discover that she's not. Again interpretation originates in obliviousness to differences that could make a difference. Interpretation in general explains the vagaries of life, the way we can't know exactly what we're getting into, changing our lives in unpredictable ways.

Whether we have anything like free will as we imagine it, aims and the interpretations that follow from them clearly depart from strict cause-and-effect predictability. Perhaps the unpredictability of interpretation explains, at least in part, the experience of free will.

If so, it would suggest a possible alternative to the two arguments that prevail in the free will debate—on the one hand that we are determinate machines; on the other hand, that we are autonomous ghosts within machines that are somehow freed causes of otherwise uncaused effects.

INTERPRETIVE BETS

About-ness has been a theme throughout this exploration. To recap briefly, the self's functional traits are interpretations *about*, representative of, or fitted to the self's circumstances. Interpretations are not cause-and-effect phenomena. They're not effects caused by, coded for, or template-copied from circumstances. Rather, interpretations are embodied means-to-ends functional traits aimed to fit circumstances with no guarantee that they will. Interpretations are fallible. They are bets that could fail.

Interpretation is guesswork. In a sense, after first selves, interpretation is largely educated guesswork, in that adaptations are interpretations built upon successful guesses accumulated over generations. And though in this evolutionary sense interpretations are educated guesses, most guesses, with rare exceptions and chiefly in humans, are very far from conscious, deliberately placed bets.

Most fundamentally, the self is the bet—the tuned synergistic couplings sustained by self-regeneration. In the minimal autogen we have a model for that first bet, the first functional about-ness to serve a self's self-regenerative aims. This bet is a constraint emergent from molecular dynamics. This constraint's fittedness is easiest to see in the autogen's self-repair of its self-protection—the repair of seeds when capsid shells are broken. This modest capacity for self-regeneration prevents failure and thereby increases the likelihood of success, though it hardly guarantees it.

The autogen's capacity for self-regeneration is a bet on how to succeed that the autogen has no awareness it is placing. It's a bet that wouldn't be placed if it weren't for the emergent constraint. In other words, the autogen wouldn't be at risk of failure to self-regenerate if it weren't for its capacity to self-regenerate. The difference between interpretive success and failure emerges with the emergent aim to self-regenerate. Only because the autogen aims to self-regenerate is that aim at risk of failing. The standard for a successful (good vs. bad) interpretation is not imposed from the outside. It emerges with selfhood. Selves emerge as unconscious gamblers in a game that emerges with them. Selves are interpreting for their lives.

Selves are also how representation enters the universe, the self's emergent dynamics re-presenting the self's circumstances. And, again, representation is different from cause and effect. This is no clockwork universe. In a clock there's only cause-and-effect mechanism, no interpretation. Anywhere that selves emerge, interpretation emerges with them.

If interpretation were simply observation, a passive reading of circumstances, it would be interpretation from outside, a "view from nowhere" in that the interpretation does not impinge upon cause and effect.

Heisenberg's uncertainty principle reveals how observation inescapably alters events at the quantum scale, but there's a much more important observer effect to be found in the self's capacity for interpretation. It's never just a bystander gazing without perturbing. Interpretation changes the self's behavior, which generates novel work in the universe, the particular things, for example, happening here on Earth today that could not have been predicted from the laws of physics and chemistry applied to conditions in the prelife universe.

ROUND ABOUT-NESS

We saw that Laplace extrapolated from Newton's deterministic laws of motion to the idea that a demon that knew the state of all atoms in the universe would be able to calculate all future states, thus demonstrating that the future is predetermined by the past.

Notice, as others have, that a problem arises when, to calculate the future, the demon makes a representative model of all events in the universe, including the events involved in making the representative model. At that point, the demon is both model and modeler. Modeling, the demon alters what is being modeled. Representing everything—including oneself representing everything—is a bit like eating one's own mouth. The cognitive scientist Douglas Hofstadter called such a situation a *strange loop*.

A strange loop is an ambiguity about what's driving what, about who's on top, who's running the show. It's where the cause is, confusingly, the effect and vise versa. Strange loops occur when something shows up at two adjacent levels in a hierarchy, in the demon's case, in the representing and what is being represented.

Strange loops are as disorienting as a hall of mirrors, representations of representations that appear until you can't distinguish between what's representing and what's being represented. It's tangled about-ness. One consequence of this ambiguity is that you can climb up and away from square one only to find yourself back to it. It's hard to know what causes what and not just because you can't track it, but because the tracking influences what's tracked. The representing is represented. The about-ness includes itself.

Strange loops arise, for example, when the overseer oversees himself, when the judge is also the contestant, when the umpire is also the player, when governors are to govern themselves, or when the act of observing is part of what's observed.

Teachers and students are in a strange loop relationship in that each is both employer and employee. You encounter a strange loop any time you're in debate with people who slide themselves up to the judge's bench to declare themselves the winners. Strange loops are present in every shaggy dog story, and in many optical illusions, most famously those of M. C. Escher.

We encountered a strange loop problem when in our brief historical survey we found Europe and Islam struggling with a double-counting problem. If the physical laws of nature and God's will both explain reality, which governs? Which is master, which is slave? Does God control physical laws, or do they control Him?

Islamic scholars escaped the double counting through faith in Allah, the ungoverned governor. Europe escaped the double counting through faith in determinism under the laws of nature, the eliminativist-leaning attempt to claim determinist cause and effect uncontrolled and supreme. Laplace's conjecture followed directly from Europe's attempt to escape double counting, yet it doesn't escape it. The demon interprets and, in so doing, influences what it is interpreting.

SELVES ARE STRANGE LOOPS

We evolved selves are strange loops also, tangled hierarchies of levels of representation. We got a glimpse of this with the template autogen—the way that functional traits became redundantly represented, expressed both in the autogen's dynamics and incipiently in static signs, constrained patterns within the template molecules. Which constrains which? They both constrain and are constrained by each other in a strange loop relationship.

Fully developed template molecules like DNA in all known species are not blueprints specifying every constraint in a body, but are more complete constraints than can be found in the template autogen. Lamarck was wrong about life experiences being heritable. Nonetheless, genes and body dynamics are a strange loop, a tangled hierarchy. Genes are not masters dictating everything about their slave bodies. The body's dynamics also influence gene expression.

Between template molecule representation and somatic (body) representation there's a strange loop. Like Laplace's demon representing itself with everything else, the mechanisms of genetics represent, along with everything else, the mechanisms of gene representation.

In this brief survey of Deacon's approach to solving the mystery of selves, I have skipped over most selves. I've concentrated on the origin of selves and its implications for us humans with hardly a word about the majority of selves.

Such glossing would be inexcusable but for the excuse that Deacon is still tentative about his bets here and I've aimed to keep this book brief. But to point in a direction that this research is taking, notice that we human selves have evolved several levels of a strange loop–tangled hierarchy. From autogenlike tuned dynamic couplings, to genes, to neuronal activity, to brains, to emotions, to language, we humans are many layers deep of strange loops accumulated on strange loops.

Douglas Hofstadter, who coined the term *strange loop* and has done magnificent work illuminating their presence and consequences, introduced the argument that selves are strange loops in his book pointedly titled *I Am a Strange Loop.*[7] Through the autogen model and further speculations, Deacon aims to detail how our strange loop nature emerged and works. Here, though, are two takeaways from even these early stages of this pioneering work.

Selfhood explains a major shift from the modest incomplete determinism of classical cause-and-effect physical dynamics to the radically incomplete determinism that starts with the emergence of selves and finds its most magnificent and maddeningly unpredictable manifestation in humans.

There are many eliminativism-leaning researchers who today talk as though it has been established that selves are just cause-and-effect machines, just fancier computers, and the only remaining question is how to break it to us.[8] Through equivocation they fail to notice the important distinction between causal one-to-one coding and interpretative representation. We are not coding machines. We are strange loop interpretive guesses, the radical imposition of emergent self-regenerative constraints that make us whole selves, emergent from less—that is, from reductions in possible molecule-on-molecule dynamics.

29

MAKING SCIENCE SAFE FOR VALUE

HUME'S GUILLOTINE

Scientists have long had a complicated relationship with value—good and bad, better and worse, moral and immoral. Early in the awakening of the scientific mind, supernatural values dominated in culture. During the Enlightenment, natural philosophers slowly divorced themselves from the constraints that supernatural values imposed, setting aside moral questions as not for scientists to arbitrate. Science would be for the description and explanation of what exists, not for how one ought to behave with what exists.

Enlightenment philosopher David Hume posited what came to be called *Hume's guillotine*, the argument that there is no way to deduce an *ought* from an *is*. Deduction is the gold standard of irrefutability. One could have a descriptive and explanatory deduction, or one could have a prescriptive (an ought) deduction, but Hume argued that there's no way to bridge deductively from one to the other. For example, returning to our deduction earlier that Socrates is mortal, there is no way to bridge from Socrates's mortality to an argument that one ought to value life.

By oughts, Hume meant universal, timeless moral absolutes, dos and don'ts that everyone should apply to all relevant circumstances. He thereby makes a point that is resonant with the perspective we have taken here. If the universe is indeed aimless and the supernatural realm, if it exists at all, has no bearing on the natural world, then it's hard to see how we could deduce any timeless, universal moral oughts from either.

The word *ought* originally meant either *possessed* or *owed*, which makes it ambiguous. A *possessed ought* would be a possessed value to a self, given its aims and circumstances, in other words, a value for selves. An *owed ought* would be an obligation beyond our selves, either to some timeless universal moral value or to other selves, given the values that they *possess*.

If an aimless universe and the supernatural don't dictate universal moral laws, then what are values and is there a reliable source for them? Deacon's approach suggests a direction that a naturalist's answer might take: what we owe to ourselves and to others originates in value, standards possessed by selves as a function of their aims and circumstances.[1]

Science is the pursuit of the most accurate and value-free descriptions of what is, and so far the scientific description of *what is* suggests a universe that shows no signs of aims in the first ten billion years. Those who embrace science would therefore be mum—within scientific discourse—on the subject of timeless, universal oughts owed to the universe, deriving from a God or a higher power. Less tactfully, our knowledge of the universe might suggest to some that such oughts don't exist.

Still, it is useful to distinguish timeless, universal moral oughts from values expressed by selves given their aims. Universal ought's and a self's values are often conflated, leaving scientists mum or at least reticent to broach value for selves. In this they would be throwing the value-baby out with the moral-absolute bathwater.

To some in the scientific community, Hume's guillotine suggests that science cannot speak to questions of value at all. The physical sciences have given scientists plenty to explore without crossing over into a discussion of value. Life and behavioral scientists would have a harder time of it, since value is intrinsic to their central concern: selves and aims. Still, many try, either through eliminativism, the argument that suggests that values aren't real, or through the treatment of values as no different from material causes of material effects, the pool player's desire *causing* the eight ball to travel toward the corner pocket.

Deacon's approach suggests that if we take *ought* to mean value for selves, we can't escape this fundamental descriptive fact: *oughts are*. Our real natural universe, the universe that scientists investigate with as much neutrality as they can muster, contains selves with values expressed through their aims. Values do exist in our universe, not as timeless universal moral absolutes imposed by the universe, but as emergent with selves.

NONOVERLAPPING MAGISTERIA

When scientists first tried to eke out room for neutral description of what is, the intellectual world was awash in supernatural ought's, moral absolutes declared to be higher than values for selves. Scientists confronted a choice about whether to try to override supernaturalism or merely to defend scientific neutral ground without trespassing on the supernaturalist moral high ground.

Paleontologist and champion of evolutionary theory Stephen Jay Gould proposed a truce with religion in what he called *nonoverlapping magisteria*, in other words, two independent domains, one for science and one for religion, arguing that, "The net of science covers the empirical universe: what is it made of (fact) and why does it work this way (theory). The net of religion extends over questions of moral meaning and value. These two magisteria do not overlap." To Gould, this truce represented "a principled position on moral and intellectual grounds, not a mere diplomatic stance."[2]

Notice the ambiguity in Gould's ceding to religion "questions of moral meaning and value." Is it moral meanings and moral values or is it moral meanings and values for selves? If Gould's is not "a mere diplomatic stance," it can suggest that science not only shouldn't but also can't speak to questions of value. Ironically, in making a scientific case that science ought not speak to value, Gould crosses his own line between separate magisteria—a scientist arguing that scientists ought not speak in terms of oughts.

Gould's view can be interpreted as consistent with a strong eliminativist position. If the universe is materially determined, if everything about selves and aims can be reduced to material cause-and-effect interactions, then there simply is no such thing as value. As such, if science were to speak to questions of value, it would pronounce value nonexistent. Scientists would counsel sterile nihilism, the universe as a computer serving no one's aims, the self as mere chemical dynamics.

Few scientists would declare nihilism outright, but people often get the impression that this is where science is taking us, to a universe in which value is an illusion at all scales—no moral absolutes and no values, since selves, like the universe, are mere cause-and-effect dynamics. By this account, life is meaningless and valueless.

Or, if not entirely valueless, of infinitesimal value. From the expansive scientific perspective that has brought us news about the whole universe, our values recede from center stage, shrinking to nothing at the vanishing point of an ever-bigger picture.

This does not sit well with most of us, which may help explain the current moral absolutist backlash and religious resurgence, at least in some parts of the world. Terror management theory research reveals that, when faced with our own mortality and therefore the relative insignificance of our selves and our aims, we tend to dig in our heels on our values, championing what psychologist Ernest Becker called our "immortality campaigns," our campaigns to assert our personal value as defenders of timeless moral absolutes.[3]

Deacon puts our predicament this way:

> Perhaps the most tragic feature of our age is that just when we have developed a truly universal perspective from which to appreciate the vastness of the cosmos, the causal complexity of material processes, and the chemical machinery of life, we have at the same time conceived the realm of value as radically alienated from this seemingly complete understanding of the fabric of existence. In the natural sciences there appears to be no place for right/wrong, meaningful/meaningless, beauty/ugliness, good/evil, love/hate, and so forth. The success of contemporary science appears to have dethroned the gods and left no foundation upon which unimpeachable values can rest. Philosophers have further supported this nihilistic conception of scientific knowledge by proclaiming that no assessment of the way things are can provide a basis for assessing how things should be. This is the ultimate heritage of the Cartesian wound that severed mind from body at the birth of modern science. The removal of any approach to value from a scientific perspective is the ultimate expression of having accepted the presumed necessity of that elective surgery.[4]

NO LONGER STUCK BETWEEN A ROCK AND A HARD PLACE

Although moral absolutism is typically associated with supernaturalism, it no longer resides there exclusively. Indeed, one could readily draw moral absolutism from natural science. Many have tried to do so from evolutionary theory.

Darwin's world is the world of evolving winners and losers. Though Darwin didn't draw moral conclusions from his work, many seem to have used it as a scientific Rorschach inkblot in which they could find affirmation for their absolutist moral convictions.

Evolutionary theory became a determinist formula for calculating the morally inevitable. Social Darwinists, Communists, nationalists of various nations, and romantics saw in evolutionary theory a proof of their absolutist moral standards—for social Darwinists, that the industrious and rich should prevail; for Communists, that the workers should prevail; for some nineteenth-century Americans, that the only good Indian was a dead Indian, that black people should always be kept below white people, and that Manifest Destiny should stretch across the Pacific to the Philippines; for some mid-twentieth century German nationalists, that the Jews should be exterminated; and for some romantics, that the omega point of evolution is higher consciousness or universal love.

Across today's divide between a supposedly scientific eliminative claim that there are no values and the fundamentalist assertion of universal values, there's enough aversion that a no-fly zone may seem in order.

Culturally, we seem to be stuck between a rock-sterile nihilism and a hard-and-fast moral absolutism, as though those were the only two options.

But Deacon's approach suggests a third option. Through the autogen, we discover that it may be possible for value to emerge from sterile dynamics. Autogen theory makes the natural sciences safe for the exploration of value—real aims emergent within an otherwise aimless universe.

As such, it suggests a way to mend a deep sociocultural rift, a way for biologists and other scientists to embrace rather than skirt issues of value and for nonscientists to feel safer with the continued progress of the scientific revolution, no longer wary that it will corner them with a material universe devoid of the very life blood of our existence. Mattering is real though not a feature of the universe. Our priorities and aims are real. They matter to us, and though they do not matter to the universe, they are natural phenomena, and must be accounted for in any true "theory of everything."

AT HOME (DETHRONED) IN THE UNIVERSE

If value emerges in something like the autogen, it emerges as value for particular selves given their aims and circumstances. Political scientists remind us that where we stand depends on where we sit; in other words, what we value depends on our circumstances. The autogen model suggests that this is true not just in political science but also across the life and social sciences.

Value is not self-defined but defined by selves—what's good and right for them, not for the universe. Nonhuman selves cannot put their values into words. Still we know a self's values by its aimed labor, at core, by the ways that self-regeneration constrains work, channeling it into maintaining self-regeneration. Selves aim interpretively. It's guesswork. A self has multiple subordinate aims, and human selves proliferate such aims, the multiple motives that make up a mind.

A self's aims are sometimes in conflict with one another, as we saw with the challenge of selective interaction. From the selective autogen onward, selves have accumulated wisdom, the ability to prioritize between their competing aims, and the wisdom to know the circumstantial differences that make an interpretive difference, for example, in the selective autogen, between when to interact and not interact, given the presence or absence of reactants. A self's aims and therefore its values are inescapably fallible. We can easily guess wrong about what to do in our circumstances.

Interpretations are possessed by selves. They are not information that the universe delivers preinterpreted. The universe doesn't tell selves anything. Rather, selves interpret potential signs.

We can try to generalize to timeless, universal values by extrapolating from what most or all selves do. We have done so here, in identifying the most fundamental value as self-regeneration. But it doesn't take us very far into the details of moral behavior. Self-regeneration suggests decidedly different values for different selves. More to the point here, we can't say that self-regeneration is inherently of value for the universe, which the ten billion years of prebiotic natural history suggest has no aims of its own.

With humans, self-regeneration can become subordinated to other aims, as when humans willingly sacrifice their lives, and therefore their self-regenerative aims. When we sacrifice our lives like this, we do it because we care about some other selves' values and aims, real or imagined.

When we sacrifice ourselves for our country or for future generations, we are demonstrating care for real selves. When we sacrifice ourselves for God, we are imagining God as a self with aims we aim to serve. Selves and aims drive value. Saving the earth is not about saving this sterile rock we live on, but the selves who inhabit it and depend on it.

THE STRUGGLE FOR INSISTENCE

People say there's no accounting for taste. There is accounting, but only when we have a solution to the mystery of purpose, and therefore an explanation for the values or tastes reflected in the self's aims. *De gustibus non est disputandum* (in matters of taste there can be no disputes) is not right either. There will be disputes. Indeed there have always been with life, as selves throughout natural history have worked to prevail in their struggle for existence. We may claim that selfishness is universally bad, but at core, every self is inherently selfish, valuing what it values in order to best flourish. And from this selfish start can come altruism and a devotion to the common good that transcends the individual self.

We humans can sometimes agree to disagree, but this can be challenging when we share in the consequences of our efforts. *Live and let live* is easiest with those we don't have to live with.

We will dispute and negotiate over relative values, especially today, given our symbolic and therefore technical competence and the unprecedented leverage it yields us. Technology gives us extraordinary influence over other selves' lives, forcing us to live more intimately with other people's values. Live and let live gets harder when we have to live with so many collective consequences. In matters of taste today, there's no escaping dispute.

Evolutionary theory suggests strongly that we get used to it, resigning ourselves to negotiated relativism between competing values. Negotiation is the way of all life: competition and cooperation, constraint and relaxation, ever evolving by trial and error toward what works or at least toward the most popular guesses as to what will. Darwin described his book, *On the Origin of Species*, as "one long argument." Life itself is one long argument.

We think of resistance to Darwin's theory as largely driven by a sense that it diminishes God's role and shows that we and the Earth's other primates share common ancestors. Perhaps the more fundamental resistance is driven by the sense that it suggests that winners and losers are inevitable, and yet there is no permanent winning formula, no timeless universal set of oughts. Given our ever-shifting, interdependent values and the ways in which we can never know all of what we're getting into, what fits today may not fit tomorrow. Just when a self lands on a winning formula, circumstances change and the winning formula loses.

This can't sit well with us. Life can feel like a major catastrophe playing out over geological time. We aim to survive. We see others around us failing to survive. We want desperately to find surefire moral solutions to the puzzle of life, the reliably timeless and universal oughts that assure survival of what we value most. Evolutionary theory suggests that there isn't one. We guess what will serve our aims and which aims to serve, but we can guess wrong.

Science itself is a commitment to this guesswork, the trial-and-error process, a dedicated embrace of "let the best solution win" negotiation. In science, no one ever gets the last word. Scientists only ever posit the best theory so far, to be beat by a better theory should one come along. Scientists can never declare the final victorious theory any more than natural selection ever hones to a final victor prevailing eternally in the struggle for existence.

In a way then, evolutionary theory already provides an alternative to nihilism and moral absolutes, but one that, lacking an explanation for the origin of evolution—the mystery of purpose, has not been grounded in the physical sciences.

Hovering mysteriously above that most solid scientific ground, the evolutionary approach is not just disappointing to the human impulse to find the surefire formula. It becomes just another ungrounded dogma in the fight.

At minimum, Deacon focuses us in the right register with the right questions, not supernatural questions, not questions of values to the universe, not origin-of-life questions that attend only to physical cause-and-effect dynamics with no focus on the emergence of selves and aims, and not moral debate that explores values as though they were sky hooked into the world with human consciousness.

Deacon focuses us on the very origin of value. He asks: Is there a testable, physically feasible model whereby selves and aims could emerge from aimless dynamics? He then suggests a potentially viable answer, one that resonates with what is obvious and intuitive:

Selves have values. Values diverge. We negotiate over values.

LIKELY AIMS IN AN APPARENTLY AIMLESS UNIVERSE

At the rate things are going on Earth, there is a risk that we will be faced with convincing evidence that human culture, quite new in the history of the universe, is a short-lived evolutionary experiment. Climate crisis, nuclear disaster, or other consequences of mature technologies in our young hands may doom civilization as we know it. This is the most unspeakable tragedy we can imagine. So tragic we can barely stand to imagine it, even as the evidence mounts.[5]

Over the coming decades, we may also gain convincing evidence of selves and aims elsewhere in the universe. And with efforts like Deacon's, we can achieve a plausible natural-science explanation for how selves and aims can emerge anywhere even if such evidence does not materialize.

These breakthrough realizations converge on a strange and cosmic consolation. Though mass extinctions really occur, selves are always possible somewhere in the universe. Selves are not just possible on faith or on the evidence from our one little planet, but possible in a way that we, in our brief experiment with exploring our condition, can come to understand.

Fortunately, the death of human civilization is not upon us yet. If by consensus there's any moral value worth promoting at the convergence of our aims, it may simply be this: the highest moral aim is keeping selves self-regenerating.

As far as we know, life doesn't mean anything to the universe. We make the meaning, and if we were to make a meaning that most of us could agree on, it might be that the meaning of life is not to end it.

Perhaps the meaning we make should also make a priority of languaged life, the life of any rare symbolic species like us, special because such a capacity is likely very rare in the universe. Symbolic competence is likely always a late-evolving trait. It makes for a precarious kind of foraging, not just for the energy and resources necessary for self-regeneration but also for self-understanding.

Symbolic life is what opens us to the vantage point from which we can wonder about the universe, modeling it, debating it, and, in the process, discovering by trial and error what it is to be alive.

ACKNOWLEDGMENTS

As I put final touches on this book that has been my focus for well over two years, I'm reminded of one of the first insights I gleaned from evolutionary theory: We didn't fall from grace, we rose from slime.

Creation is the trial-and-error organization of the disorganized. Disorganization is the foundation. My terrible first draft and the next umpteenth drafts were indeed failures, but not because I had fallen out of excellence—rather, because I was scrambling to find it, much of it through a process of elimination, constraining the inherently unconstrained. I easily eliminated six hundred pages to arrive at these remaining pages.

As my text came into focus, the illusion that one falls from grace returned. I thought, "Well, why didn't I just say it right the first time?" And my answer: Because I've wasted my whole life learning things I now already know.

My first gratitude therefore goes to the great chain of thinking, the researchers past and present who dedicated their lives to learning things we now already know, their research contributing to the trial-and-error process by which we attempt to discover our place in the cosmos. These would include the researchers cited in this book and countless more. If we each had to learn everything from a cold start, we would never get anywhere. Thanks memory—both individual and collective.

And to future researchers, who will refine, correct, and update the trials floated in this book and elsewhere, gratitude for their forging forward to things that future generations will already know.

My deepest personal gratitude is to Terrence Deacon, my mentor, colleague, role model, and friend for twenty years, the most patient, assiduous, and unassuming brilliance I've ever known. Deacon seems tone-deaf to questions of status. He doesn't pull rank, toot his horn, or rest a self-serving finger on the scale. He seems immune to the "not invented here" syndrome, grateful and acknowledging of good ideas from any source, and calmly dismissive even of his own bad ideas.

His is a patient and deliberate struggle to bring order to the inherent disorder of thought. His appreciations and discouragements are always mild, precise, and tentative, conjectures that encourage his students and colleagues to continue to think it up yet another notch collaboratively, free from fear of losing, or hopes of gaining status.

About a decade ago, a friend dubbed Terry's colleagues "the pirates," a reference to the post–World War II comic strip *Terry and the Pirates*. The pirates are a loose gathering of people who meet weekly at Terry's house to continue refining the theories presented here. Terry's wife, Ella Ray Deacon, has been a friend, guide, collaborator, and ally throughout. I'm grateful for her hospitality and wisdom.

I'm also grateful to fellow pirates Zena Kruzick, Auguste Nahas, Jay Ogilvy, Tom Rockwell, Maximus Thaler, and Adrian Wyard, who critiqued and edited my full manuscript, some of them patiently wading through multiple drafts.

I'm grateful to the pirates who reviewed sections relevant to their areas of specialization or interest, including Mark Carranza, Ty Cashman, Zann Gill, Mani Hamidi, Julie Hui, Stefan Leijnen, Spyridon Koutroufinis, Joanna Leonardi, José Monserrat Neto, Felipe Veloso, and Hajime Yamauchi.

And to other pirates for their insights over the years: Joshua Bacigalupi, Diego Caleiro, Harvey Firestone, James Haag, Aran Gharibpor, Mark Graves, Eduardo Kohn, Bob Logan, Ambrose Nahas, Felipe-Andrés Piedra, and Rodrigo Vanegas.

I'm grateful indeed to Wendy Lochner and Christine Dunbar at Columbia University Press; to my peer reviewers, Ursula Goodenough and Loyal Rue; to my editors, Cecelia Cancelaro, Beth Chapple, Robert Demke, and Hilary Hinzmann; and to outside interested readers, including Peter Barnes, Bill Carey, Maria Ciccone, Teddy Shalon, and Gary Zellerbach.

Finally, my gratitude goes out to my family, social circle, former students, and *Psychology Today* blog readers, who encourage me just enough, sometimes just through sparkly accommodation of my eccentric priorities. These include my children, Lucy, Alex, and Will Sherman, and my eldest brother, Stuart Sherman, as well as Kaja Perina and Hara Marano at *Psychology Today*.

And to the many muses who come and go, whose interesting, curious lives have nourished me with food for thought.

APPENDIX

THE FRAMEWORK AT A GLANCE

The mystery of purpose has been explored under a wide range of frameworks. Here is the framework I have employed, presented as a step-by-step (nonalphabetical) collection of definitions, some of them divergent from convention, as with my use of *selves* broadened to include all organisms. In parenthesis (IN:) are Deacon's technical terms as employed in his book *Incomplete Nature: How Mind Emerges from Matter.*

PLACING IN PHILOSOPHICAL CONTEXT

ONTOLOGY: The exploration of what exists. *An* ontology is a theory about what exists.

SEMIOTICS: The exploration of how selves interpret signs as significant, sensing and responding in ways that matter for the selves. *A semiotic* is a theory of how selves interpret.

INTERPRETATION: The definition broadened here to include all of the ways that selves purposefully sense and respond to their circumstances. Interpretation is work done by and for selves about their circumstances given their aims. All functional traits and behaviors are interpretations.

Epistemology: The philosophy of knowledge—what it is and how it's acquired, here broadened to include all varieties of fittedness of a self to its circumstances. By this definition, traits and behaviors, even strictly biological ones that enable a self to succeed, are a kind of knowledge.

Teleology: The exploration of what purpose is and how it works. A *teleology* is a particular theory of what purpose is and how it works.

Purpose (Greek, *Telos*): The definition broadened to include all aims, whether generated "on purpose" (consciously, deliberately, with intention) or not. Purpose thus includes all functional means-to-ends traits and behaviors in all selves.

Teleologies: A *supernatural teleology* assumes that purpose comes from outside of nature. A *natural teleology* assumes that purpose must be explained as part of nature (for example, *pansychism*, a strain of natural teleology, treats nature as imbued with purpose, for example, that even subatomic particles have aims). An *emergent teleology* is a kind of natural teleology that attempts to explain how purposes emerge with selves. Deacon's solution is *an emergent teleology.*

The Mystery of Purpose: The question "What are selves and aims and how could they emerge within a universe that is otherwise aimless?" Solving the mystery would explain the *ontology of semiotics*, the real existence of interpretation anywhere it exists in the natural universe. Embracing the mystery, one assumes that selves interpreting circumstances to serve their aims haven't always existed. Their existence must be explained as emergent from cause-and-effect physical chemistry.

Deacon's Solution: A proposed theory that explains the emergent ontology of semiotics, epistemology, and teleology, the existence of interpretation, fittedness, and purpose emergent with selves and their aims.

EXISTENT ELEMENTS TO BE EXPLAINED

Selves: Entities capable of aiming, doing self-directed work, self-regenerative work by and for themselves given their circumstances. Nonselves include dead bodies, machines, and all the inanimate dynamics in the universe. Selves include all living individuals ever, anywhere in the universe, including preevolutionary selves.

PREEVOLUTIONARY SELVES (EMERGENT SELVES, FIRST SELVES, MISSING LINKS): Selves existent prior to natural selection. Natural selection is differential survival of selves (not just differential replication of chemicals). Thus for natural selection to operate, selves must already exist.

AIMS (IN: ENTENTIONAL BEHAVIOR): Constraints upon possible work such that self-directed, self-regenerative work is achieved.

SELF-DIRECTED WORK: Work constrained to what enables a self to continue to function. Self-directed work is inherently circular, work by and for a self, work to keep working or more specifically work that prevents the failure of a self's ability to keep working. Self-directed work is necessary to outpace the second law of thermodynamics, the universal tendency for things to stop working.

A SOLUTION TO THE MYSTERY OF PURPOSE WOULD EXPLAIN THE ORIGIN AND NATURE OF THE FOLLOWING

SELF-REGENERATION (IN: TELEODYNAMICS): A self's defining aim. Self-regeneration is what selves do and what nonselves don't. Self-regeneration entails three primary capabilities:

1. Self-Repair: The capacity to regenerate regularities faster than they would otherwise degenerate, given the second law tendency toward irregularity.
2. Self-Protection: The prevention of degeneration through any features that enable a self to resist the tendency to degenerate given the second law, by segregating dynamics, for example, a self's protective cell walls, shell, or skin that contains protoplasm, preventing it from diffusing and dispersing as it would otherwise do, given the second law.
3. Self-Reproduction: The capacity to proliferate varied offspring that also have the capacity for self-regeneration.

SELECTIVE INTERACTION: A challenge resulting from the conflicting requirements of self-regeneration: The requirement for interaction with the environment to acquire resources and energy for self-repair and self-reproduction, while maintaining self-protection from deleterious interaction.

In other words, being open to what facilitates but not what impedes self-regeneration.

FOR-NESS: Something being of value or significance, ultimately for a self. Nothing is of value or significant for inanimate things—not even for machines.

ABOUT-NESS: The property of referring to, representing, or being organized with respect to something else that is of value or significance for a self *with respect to the self's circumstances or environment.*

FUNCTIONAL TRAIT TRIAD: Three differences linked by for-ness and aboutness: (a difference in X is of value or significance) *for* (a Self) *about* (fulfilling its aims in its circumstances).

BARRED ASSUMPTIONS

SUPERNATURALISM: The assumption that the solution to the mystery of purpose resides in a realm inaccessible to science.

EQUIVOCATION (IN: CRYPTIC DUALISM): The tendency to treat phenomenon interchangeably as purposeful and nonpurposeful.

THRESHOLDISM: The assumption that once cause-and-effect interactions cross a threshold level of complexity they become purposeful selves with aims.

ELIMINATIVISM: The assumption that purpose need not be explained because it doesn't really exist.

REVERSE-ENGINEER FALLACY: The functionalist assumption that the mechanisms we would engineer into a device to make it function like selves illuminates how selves really function.

AMNESIC WATCHMAKER SYNDROME: The habit of ignoring the engineering involved in producing a functional machine when implying that machines explain emergent selves.

SACRIFICING VIABILITY FOR INVIOLABILITY: The tendency to concentrate on the tractable while ignoring the intractable aspects of purposeful behavior as a means to demonstrate that purposeful behavior is entirely tractable mechanism.

DEACON'S SOLUTION TO THE MYSTERY

THE SECOND LAW (IN: HOMEODYNAMICS): The universal tendency for segregated regularities when allowed to interact to become irregular, in other words for order to become disordered and organization to become random.

IRREGULARITY: Disorder, randomness, mixed-upness, the opposite of regularity.

REGULARITY: Repetition, redundancy, repeated pattern. Regularity is a less ambiguous name for what is more abstractly called order or organization.

DYNAMICS: The energetic interaction of large quantities of elements.

DYNAMIC PATHS (OR JUST PATHS): The spatiotemporal ways that dynamics could fall toward second law irregularity.

CURRENTS: Paths taken.

DYNAMIC TENDENCIES (IN: ORTHOGRADES): Probable currents, the paths that dynamics are likely to take.

DYNAMIC COUNTERTENDENCIES (IN: CONTRAGRADES): Currents impeding one another. Currents becoming impasses to one another, thereby changing probable dynamic tendencies.

EMERGENCE: The remaining paths that result from dynamic countertendencies.

ENERGY: An energetic difference equalizing due to the second law, segregated energetic regularities becoming irregular (mixed up) when allowed to interact.

CONSTRAINT: Any thing or process that reduces the likelihood of dynamics taking paths.

IMPOSED CONSTRAINT: Reduction of paths imposed upon dynamics from sources other than the dynamics, for example, a riverbed constraining water flow to some paths and not others.

EMERGENT CONSTRAINT: Constrained paths arising throughout dynamic interaction as currents impede one another, making some paths less likely, thereby making remaining paths more likely.

Emergent Regularization (IN: Morphodynamics): A less ambiguous term than *self-organization*. Emergent regularization is one of two kinds of emergent constraint. Examples discussed include whirlpools, Bénard cells, autocatalysis, crystal (for example, capsid) formation.

Coupling: Two or more emergent regularization processes interacting.

Synergy: Not a whole that is more than the sum of its parts but a whole emergent from dynamic countertendencies impeding one another such that less than the sum of possible currents are present.

Synergistic Coupling: Dynamic countertendencies between two or more emergent regularization dynamics such that synergy is achieved.

Emergent Self-Regeneration (IN: Teleodynamics): The second kind of emergent constraint necessary for the emergence of self-regeneration and therefore for selves and aims. Emergent self-regenerations are a further constraint resulting from the synergistic coupling of two or more underlying emergent regularization dynamics.

DEACON BIBLIOGRAPHY

JOURNAL ARTICLES

Deacon, Terrence, and Tyrone Cashman. "Steps to a Metaphysics of Incompleteness." *Theology and Science* 14 (2016).

Deacon, Terrence. "Reconsidering Darwin's 'Several Powers.'" *Biosemiotics* 9 (2016): 121–128.

Deacon, Terrence. "Steps to a Science of Biosemiotics." *Green Letters* 19 (2016): 293–311.

Deacon, Terrence, Alok Srivastava, and Joshua Bacigalupi. "The Transition from Constraint to Regulation at the Origin of Life." *Frontiers in Bioscience* 19 (2014): 945–957.

Deacon, Terrence, and Spyridon Koutroufinis. "Complexity and Dynamic Depth." *Information* 5 (2014): 404–423. doi:10.3390/info5030404 (ISSN 2078–2489).

Deacon, Terrence. "Making Sense of Incompleteness: A Response to Commentaries." Paper delivered at book symposium on Terrence Deacon's *Incomplete Nature: Religion, Brain and Behavior*, 2013.

Deacon, Terrence. "The Importance of What's Missing." *New Scientist*, November 2011, 36–42.

Deacon, Terrence. "A Role for Relaxed Selection in the Evolution of the Language Capacity." *Proceedings of the National Academy of Sciences* 107 (2010): 9000–9006.

Kull, Kalevi, Terrence Deacon, Claude Emmeche, Jesper Hoffmeyer, and Frederik Stjernfelt. "Theses on Biosemiotics: Prolegomena to a Theoretical Biology." *Biological Theory* 4, no. 2 (2009): 167–173.

Hui, Julie, and Terrence Deacon. "The Evolution of Altruism via Social Addiction." *Proceedings of the British Academy* 158 (2009): 181–203.

Deacon, Terrence, and Tyrone Cashman. "The Role of Symbolic Capacity in the Origins of Religion." *Journal of Religion, Nature and Culture* 3, no. 4 (2009): 490–517.

Deacon, Terrence. "Shannon-Boltzmann-Darwin: Redefining Information. Part 1." *Cognitive Semiotics* 1 (2007): 123–148.

Deacon, Terrence. "Shannon-Boltzmann-Darwin: Redefining Information. Part 2." *Cognitive Semiotics* 2 (2008): 167–194.

Sherman, Jeremy, and Terrence Deacon. "Teleology for the Perplexed: How Matter Began to Matter." *Zygon* 42 (2007): 873–901.

Deacon, Terrence. "Reciprocal Linkage Between Self-Organizing Processes Is Sufficient for Self-Reproduction and Evolvability." *Biological Theory* 1, no. 2 (2006): 136–149.

Goodenough, Ursula, and Terrence Deacon. "From Biology to Consciousness to Morality." *Zygon* 38 (2003): 801–819.

BOOKS

Schilhab, Theresa, Frederik Stjernfeldt, and Terrence Deacon, eds. *The Symbolic Species Evolved.* Dordrecht: Springer, 2012.

Deacon, Terrence. *Incomplete Nature: How Mind Emerged from Matter.* New York: Norton, 2012.

Deacon, Terrence. *The Symbolic Species: The Co-Evolution of Language and the Brain.* New York: Norton, 1997.

BOOK CHAPTERS

Leijnen, Stefan, Tom Heskes, and Terrence Deacon. "Exploring Constraint: Simulating Self-Organization and Autogenesis in the Autogenic Automaton." In *Proceedings of the Artificial Life Conference 2016*, edited by Carlos Gershenson, Tom Froese, Jesus M. Siqueiros, Wendy Aguilar, Eduardo J. Izquierdo, and Hiroki Sayama. Cambridge: MIT Press, 2016.

Deacon, Terrence. "Information and Reference." In *Representation and Reality: Humans, Animals and Machines*, edited by Gordana Dodig-Crnkovic and Raffaela Giovagnoli. Heidelberg: Springer, 2016.

Deacon, Terrence. "Towards a General Theory of Evolution." In *Living Machines: A Handbook of Research in Biomimetic and Biohybrid Systems*, edited by Tony J. Prescott and Paul F. M. J. Verschure. Heidelberg: Springer, 2016.

Deacon, Terrence. "The Emergent Process of Thinking as Reflected in Language Processing." In *Thinking: First Person in Interaction: Pragmatism, Phenomenology and Psychotherapy*, edited by Donata Schoeller and Vera Saller. Freiburg: Alber, 2016.

Deacon, Terrence. "Semiosis: From Taxonomy to Process." In *Charles Sanders Peirce in His Own Words*, edited by Torkild Thellefsen and Bent Sørensen. Berlin: De Gruyter, 2014.

Deacon, Terrence, and Tyrone Cashman. "Teleology Versus Mechanism in Biology: Beyond Self-Organization." In *Beyond Mechanism: Putting Life Back Into Biology*, edited by Brian G. Henning and Adam C. Scarfe, 287–308. Lanham, MD: Lexington, 2012.

Deacon, Terrence. "Information." In *A More Developed Sign*, edited by Donald Favareau, Paul Cobley, and Kalevi Kull, 161–164. Tartu, Estonia: Tartu University Press, 2012.

Deacon, Terrence. "Beyond The Symbolic Species." In *The Symbolic Species Evolved*, edited by Theresa Schilhab, Frederik Stjernfeldt, and Terrence Deacon, 9–38. Dordrecht: Springer, 2012.

Deacon, Terrence. "Darwin's 'Several Powers.'" In *Biosemiotics Turning Wild*, edited by Timo Maran, Kati Lindström, Riin Magnus, and Morten Tønnessen, 71–78. Tartu, Estonia: Tartu University Press, 2012.

Deacon, Terrence. "The Symbol Concept." In *The Oxford Handbook of Language Evolution*, edited by Maggie Tallerman and Kathleen R. Gibson, 393–405. Oxford: Oxford University Press, 2011.

Deacon, Terrence, and Tyrone Cashman. "Eliminativism, Complexity and Emergence." In *The Routledge Companion to Religion and Science*, edited by James Haag, Gregory Peterson, and Michael Spezio, 193–206. New York: Routledge, 2012.

Deacon, Terrence, James Haag, and Jay Ogilvy. "The Emergence of Self." In *In Search of Self: Interdisciplinary Perspectives on Personhood*, edited by J. Wentzel van Huyssteen and Erik P. Wiebe. Grand Rapids: Eerdmans, 2011.

Deacon, Terrence. "What's Missing from Theories of Information?" In *Information and the Nature of Reality: From Physics to Metaphysics*, edited by Paul Davies and Niels Henrik Gregersen, 146–169. New York: Cambridge University Press, 2010.

Deacon, Terrence, and Jeremy Sherman. "The Pattern Which Connects Pleroma to Creatura: The Autocell Bridge from Physics to Life." In *A Legacy for Living Systems*, edited by Jesper Hoffmeyer, 59–76. Dordrecht: Springer, 2008.

Hui, Julie, Tyrone Cashman, and Terrence Deacon. "Bateson's Method: Double Description: What Is It? How Does It Work? What Do We Learn?" In *A Legacy for Living Systems*, edited by Jesper Hoffmeyer, 77–92. Dordrecht: Springer, 2008.

Deacon, Terrence, and Jeremy Sherman. "The Physical Origins of Purposive Systems." In *Embodiment in Cognition and Culture*, edited by John Michael Krois, Mats Rosengren, Angela Steidele, and Dirk Westerkamp. Amsterdam: Benjamins, 2007.

Deacon, Terrence. "Three Levels of Emergent Phenomena." In *Evolution and Emergence: Systems, Organisms, Persons*, edited by Nancey Murphy and William Stoeger, 88–112. Oxford: Oxford University Press, 2007.

Deacon, Terrence. "Emergence: The Hole at the Wheel's Hub." In *The Re-Emergence of Emergence*, edited by Philip Clayton and Paul Davies, 111–150. Cambridge: MIT Press, 2006.

Goodenough, Ursula, and Terrence Deacon. "The Sacred Emergence of Nature." In *The Oxford Handbook of Religion and Science*, edited by Philip Clayton and Zachary Simpson. Oxford: Oxford University Press, 2006.

Deacon, Terrence. "The Hierarchic Logic of Emergence: Untangling the Interdependence of Evolution and Self-Organization." In *Evolution and Learning: The Baldwin Effect Reconsidered*, edited by Bruce H. Weber and David J. Depew, 273–308. Cambridge: MIT Press, 2003.

NOTES

1. THE MYSTERY OF PURPOSE

1. Jessica Riskin, *The Restless Clock: A History of the Centuries-Long Argument Over What Makes Living Things Tick* (Chicago: University of Chicago Press, 2015), Kindle ed.
2. Ibid.
3. Addy Pross, *What Is Life? How Chemistry Becomes Biology* (Oxford: Oxford University Press, 2012), Kindle ed.
4. Ibid.
5. Sean Carroll, *The Big Picture: On the Origins of Life, Meaning, and the Universe Itself* (New York: Penguin, 2016), Kindle ed.
6. Ibid.
7. Ibid.
8. Ibid.
9. Terrence Deacon, *Symbolic Species: The Co-Evolution of Language and the Brain* (New York: Norton, 1997).

2. THE BIGGEST MYSTERY WE EVER IGNORE

1. Deacon's translation of a quotation from Ilya Prigogine and Isabelle Stengers, *La nouvelle alliance: Métamorphose de la science* (Paris: Gallimard, 1979), 278.
2. Francis Crick, *The Astonishing Hypothesis: The Scientific Search for the Soul* (New York: Scribner, 1995), 3.
3. James Watson with Andrew Berry, *DNA: The Secret of Life* (New York: Random House, 2003), 211.
4. Patricia Churchland, interview by Richard Marshall, *3:AM Magazine*, www.3ammagazine.com/3am/casual-machines/.

5. Aristotle, *Aristotle in 23 Volumes*, vols.17 and 18, trans. Hugh Tredennick (Cambridge: Harvard University Press, 1989), 438.

3. DEACON'S SOLUTION IN BRIEF

1. W. Ross Ashby, "Principles of the Self-Organizing Dynamic System," *Journal of General Psychology* 37 (1947): 125–128.
2. For example, see Alicia Juarrero, *Dynamics in Action: Intentional Behavior as a Complex System* (Cambridge, MA: Bradford, 2002).

5. SELVES

1. Terrence Deacon, *Symbolic Species: The Co-Evolution of Language and the Brain* (New York: Norton, 1997).
2. Terrence Deacon, *Incomplete Nature: How Mind Emerged from Matter* (New York: Norton, 2012), 466.
3. Deacon, *Symbolic Species*, 2.
4. René Descartes, *Discourse on the Method of Rightly Conducting the Reason*, in *Great Books of the Western World*, vol. 31, ed. Robert Maynard Hutchins (Chicago: Encyclopaedia Britannica, 1952), 60.
5. René Descartes, *Second Meditation* (2.8) and *Principia Philosophiae* (2.002), in *Great Books of the Western World*, vol. 31, ed. Robert Maynard Hutchins (1637; Chicago: Encyclopaedia Britannica, 1952), 117–118.
6. Gilbert Ryle, *The Concept of Mind* (1949; Chicago: University of Chicago Press, 2002), 11.
7. Gregory Bateson, *Mind and Nature: A Necessary Unity* (New York: Dutton, 1979).
8. Daniel Dennett, "The Origin of Selves," *Cogito* 3 (Autumn 1989): 163–173.

6. TWO GHOSTS, TWO MACHINES

1. Daniel Dennett, *Brainstorms* (Cambridge: MIT Press, 1978), 12.
2. Maynard Smith, "The Concept of Information in Biology," *Philosophy of Science* 67:177–194; as well as the reply to commentaries on pp. 214–218.
3. Colin Pittendrigh, "Adaptation, Natural Selection, and Behavior," in *Behavior and Evolution*, ed. Anne Roe and George Gaylord Simpson (New Haven: Yale University Press, 1958), 394.
4. Ernst Mayr, "Cause and Effect in Biology," in *Cause and Effect*, ed. Daniel Lerner (New York: Free Press, 1965), 33–50.
5. Ernst Mayr, "Teleological and Teleonomic: A New Analysis," *Boston Studies in the Philosophy of Science* 14 (1974): 133–159.
6. David Hull, "Philosophy and Biology," in *Philosophy of Science: Contemporary Philosophy: A New Survey*, ed. Guttorm Fløistad (Hague: Nijhoff, 1982), 2:280–316.

7. INTERPRETATION

1. Jesper Hoffmeyer, "The Natural History of Intentionality: A Biosemiotic Approach," in *The Symbolic Species Evolved*, ed. Theresa Schilhab, Frederik Stjernfelt, and Terrence Deacon (New York: Springer, 2012), 103.
2. Terrence Deacon, personal email communication with author, November 16, 2004.

8. AIMS

1. Pierre-Simon de Laplace, *A Philosophical Essay on Probabilities*, 6th ed., trans. F. W. Truscott and F. L. Emory (1814; New York: Dover, 1951), 4.
2. James Clerk Maxwell, letter to Lewis Campbell, circa July 1850, in *The Scientific Letters and Papers of James Clerk Maxwell: 1846–1862*, ed. P. M. Harman (Cambridge: Cambridge University Press 1990), 197.

9. EVOLUTION'S LIMITED LIMITING ROLE

1. James Mark Baldwin, "A New Factor in Evolution," *American Naturalist* 30 (1896): 441–451, section 5.
2. Charles Darwin, *On the Origin of Species by Means of Natural Selection or the Preservation of Favored Races in the Struggle for Life* (1859; London: Dent, 1971), 490.
3. Richard Owens, letter to the *Athenaeum* in 1863, quoted in *A Brief History of Creation*, ed. Bill Mesler and James Cleaves II (New York: Norton, 2015), 112.
4. Charles Darwin, letter to the *Athenaeum* in 1863, quoted in Mesler and Cleaves, *A Brief History of Creation*, 113.
5. Charles Darwin, *The Origin of Species*, Harvard Classics (New York: Collier, 2011), 376.

10. THE HISTORY

1. For a thorough survey of the intellectual history of debate about living agency: Jessica Riskin, *The Restless Clock: A History of the Centuries-Long Argument Over What Makes Living Things Tick* (Chicago: University of Chicago Press, 2015), Kindle ed.
2. Aristotle, *Posterior Analytics*, in *Great Books of the Western World*, vol. 8, ed. Robert Maynard Hutchins (Chicago: Encyclopaedia Britannica, 1952), 128.
3. Aristotle, *Aristotle's Politics*, in *Great Books of the Western World*, vol. 8, ed. Robert Maynard Hutchins (Chicago: Encyclopaedia Britannica, 1952), 445.
4. Al-Ghazali, *The Incoherence of the Philosophers*, trans. Michael Mamura (Salt Lake City: Brigham Young University, 1998), 363.
5. Lucretius, *The Nature of Things*, trans. Alicia Stallings (New York: Penguin, 2007). See also Stuart Gillespie and Philip Hardie, eds., *The Cambridge Companion to Lucretius*

(Cambridge: Cambridge University Press, 2007); and Steven Greenblatt, *The Swerve: How the World Became Modern* (New York: Norton, 2012).

6. Francis Bacon, *The Advancement of Learning*, in *Great Books of the Western World*, vol. 30, ed. Robert Maynard Hutchins (Chicago: Encyclopaedia Britannica, 1952), 45.
7. Benedictus de Spinoza, *A Spinoza Reader: The Ethics, and Other Works*, ed. E. M. Curley (Princeton: Princeton University Press, 1994).
8. Henry Drummond, *The Ascent of Man* (Radford, VA: Wilder, 2011), 333.

11. EVOLUTIONARY THEORY'S ELUSIVE SELF

1. Charles Darwin, *On the Origin of Species by Means of Natural Selection or the Preservation of Favored Races in the Struggle for Life*, vol. 11 (1859; London: Dent, 1971).
2. Lloyd Morgan, *The Interpretation of Nature* (Charleston, SC: Nabu, 2010).
3. Donald T. Campbell, "Blind Variation and Selective Retention in Creative Thought as in Other Knowledge Processes," *Psychological Review* 67 (1960): 380–400.
4. George Williams, *Adaptation and Natural Selection* (Princeton: Princeton University Press, 1996).
5. Richard Dawkins, *The Selfish Gene*, 30th anniversary ed. (Oxford: Oxford University Press, 2006), 254.
6. Ibid., xxi.
7. Ibid., 62.
8. Ibid, ix.
9. Ibid., xiv.
10. Ibid., 353.
11. Richard Dawkins, *The Blind Watchmaker* (New York: Norton, 1986).

12. INFORMATION ABOUT NOTHING FOR ANYONE

1. For example, the narrator in "Some of the Things That Molecules Do," episode 2 of *Cosmos: A Spacetime Odyssey* (Washington, DC: PBS, 2014), said: "DNA contains the letters of the genetic alphabet. Particular arrangements of those letters spell out instructions for all living things telling them how to grow, move, digest, sense their environment, heal, reproduce. The DNA double helix is a molecular machine."
2. Claude Shannon, "A Mathematical Theory of Communication," *Bell System Technical Journal* 27 (1948): 379–423, 623–656.
3. Ibid., 379.
4. Claude Shannon, *Collected Papers*, ed. Neil Sloane and Aaron Wyner (New York: IEEE, 1993), 180.
5. Warren Weaver, "The Mathematics of Communication," *Scientific American* (July 1949): 12.

6. Newton as quoted in Thomas Levenson, *Newton and the Counterfeiter: The Unknown Detective Career of the World's Greatest Scientist* (New York: Houghton Mifflin Harcourt, 2012), 271.
7. John Collier, "Hierarchical Dynamical Information Systems with a Focus on Biology," *Entropy* 5 (2003): 121.

13. THE ENGINEERED GHOSTS IN OUR MACHINES

1. Giulio Tononi and Christof Koch, "Consciousness: Here, There and Everywhere?," *Philosophical Transactions of the Royal Society London B* (March 30, 2015): doi:10.1098/rstb.2014.0167.
2. Thomas Hobbes, *Leviathan* (Cambridge: Harvard University Press, 1909), 14.
3. Charles Taylor, "'Fleshing Out' Artificial Life II," in *Artificial Life*, ed. Christopher G. Langton, Charles Taylor, J. Doyne Farmer, and Steen Rasmussen (Reading, MA: Addison-Wesley, 1992), 25–38.
4. Richard Lewontin, "Biology as Engineering," in *Integrative Approaches to Molecular Biology*, ed. Julio Collado-Vides, Boris Magasanik, and Temple F. Smith (Cambridge: MIT Press, 1996), 1–10.
5. Martin Gardner, "Mathematical Games: The Fantastic Combinations of John Conway's New Solitaire Game 'Life,'" *Scientific American* 223 (1970): 120–123.

14. SMALL IS DUBIOUS

1. Sean Carroll, *The Big Picture: On the Origins of Life, Meaning, and the Universe Itself* (New York: Penguin, 2016), Kindle ed.
2. Alfred Einstein, Boris Podolsky, and Nathan Rosen, "Can Quantum-Mechanical Description of Physical Reality Be Considered Complete?," *Physical Review* 47, no. 10 (1935): 777–780.
3. John Wheeler, quoted in James Gleick, *The Information: A History, a Theory, a Flood* (New York: Pantheon, 2011).
4. Terrence Deacon, *Incomplete Nature: How Mind Emerged from Matter* (New York: Norton, 2012), 74.
5. Joseph Goguen, "Towards a Social, Ethical Theory of Information," in *Social Science Research, Technical Systems and Cooperative Work: Beyond the Great Divide*, ed. Geoffrey Bowker, Les Gasser, Leigh Star, and William Turner (London: Erlbaum, 1997), 27–56.
6. Thomas Nagel, *Mind and Cosmos: Why the Materialist Neo-Darwinian Conception of Nature Is Almost Certainly False* (Oxford: Oxford University Press, 2012), 123.
7. For more about multiversalism, see Brian Greene, *The Hidden Reality: Parallel Universes and the Deep Laws of the Cosmos* (New York: Vintage, 2011).
8. Jerry Fodor, *Concepts: Where Cognitive Science Went Wrong* (New York: Clarendon, 1998), 161.

15. PROCESSES OF EMERGENT ELIMINATION

1. W. Ross Ashby, "Principles of the Self-Organizing System," in *Principles of Self-Organization: Transactions of the University of Illinois Symposium*, ed. H. Von Foerster and G. W. Zopf Jr. (London: Pergamon, 1962), 259.
2. Ibid., 273.

16. SECOND LAW IRREGULARITY

1. Terrence Deacon and Spyridon Koutroufinis, "Complexity and Dynamic Depth," *Information* 5 (2014): 404–423, doi:10.3390/info5030404 (ISSN 2078–2489).

17. EMERGENT REGULARIZATION

1. Jaegwon Kim, "Multiple Realization and the Metaphysics of Reduction," *Philosophy and Phenomenological Research* 52, no. 1 (1992): 18.

18. EMERGENT REGULARIZATION VS. EMERGENT SELF-REGENERATION

1. Rod Swenson and Michael Turvey, "Thermodynamic Reasons for Perception-Action Cycles," *Ecological Psychology* 3, no. 4 (1991): 26.
2. Erwin Schrödinger, *What Is Life? The Physical Aspect of the Living Cell* (Cambridge: Cambridge University Press, 1944), 6.
3. Terrence Deacon, Alok Srivastava, and Joshua Bacigalupi, "The Transition from Constraint to Regulation at the Origin of Life," *Frontiers in Bioscience* 19 (2014): 945–957.

20. COUPLED REGULARIZATION PROCESSES

1. Immanuel Kant, *The Critique of Judgment*, vol. 2, *Teleological Judgment*, trans. James Creed Meridith (1790; Chicago: University of Chicago Press, 1952), 321.
2. Immanuel Kant, *Critique of Judgment*, trans. J. H. Bernard (1790; New York: Hafner, 1951), 7.
3. Manfred Eigen and Peter Schuster, "Part A: Emergence of the Hypercycle," *Naturwissenschaften* 65 (1978): 7–41.
4. Pier Luigi Luisi, *The Emergence of Life: From Chemical Origins to Synthetic Biology* (Cambridge: Cambridge University Press, 2006), 159.
5. Evan Thompson, *Mind in Life: Biology, Phenomenology and the Sciences of Mind* (Cambridge: Belknap Press of Harvard University Press, 2007), 130.
6. Ibid., 115.
7. Luisi, *Emergence of Life*, 128.

8. Cited in Thompson, *Mind in Life*, 10.
9. For a more thorough critique of autopoietic units, see Terrence Deacon and Tyrone Cashman, "Teleology Versus Mechanism in Biology: Beyond Self-Organization," in *Beyond Mechanism: Putting Life Back Into Biology*, ed. Brian G. Henning and Adam C. Scarfe (Lanham, MD: Lexington, 2013).

21. AUTOGENS

1. Christopher Langton, "Computation at the Edge of Chaos," *Physica D* (1990): 42.

22. EVOLVED AUTOGENS

1. Paul G. Higgs and Niles Lehman, "The RNA World: Molecular Cooperation at the Origins of Life," *Nature Reviews Genetics* 16 (2015): 7–17, doi:10.1038/nrg3841.

24. THE CONSEQUENCES OF SELF-REGENERATION

1. Stuart Kauffman, *Investigations* (Oxford: Oxford University Press, 2000), 111.
2. Addy Pross, *What Is Life? How Chemistry Becomes Biology* (Oxford: Oxford University Press, 2012), 187.

25. CODES, SIGNS, INTERPRETERS

1. George Williams, *Adaptation and Natural Selection: A Critique of Some Current Evolutionary Thought* (Princeton: Princeton University Press, 1992), 11.
2. Richard Dawkins, *The Selfish Gene*, 30th anniversary ed. (Oxford: Oxford University Press, 2006), xxi.
3. Norbert Wiener, *The Human Use of Human Beings* (New York: Houghton Mifflin, 1954), 96.
4. Howard Pattee, "The Physics of Symbols: Bridging the Epistemic Cut," *Biosystems* 60 (2001): 5–21.

26. KINDS OF SIGNS

1. Terrence Deacon, *The Symbolic Species: The Coevolution of Language and the Brain* (New York: Norton, 1997).
2. Terrence Deacon, "Semiotics and Cybernetics: The Relevance of C. S. Peirce," unpublished manuscript, 1976.
3. Gregory Bateson, *Steps to an Ecology of Mind: Collected Essays in Anthropology, Psychiatry, Evolution, and Epistemology* (Chicago: University of Chicago Press, 1973), 428.

4. Thomas Bayes, "An Essay Toward Solving a Problem in the Doctrine of Chances," *Philosophical Transactions of the Royal Society of London* 53 (1763): 370–418, doi:10.1098/rstl.1763.0053, http://rstl.royalsocietypublishing.org/content/53.toc.
5. Terrence Deacon, "Shannon-Botzmann-Darwin: Redefining Information. Part 1," *Cognitive Semiotics* 1 (2007): 123–148; Terrence Deacon, "Shannon-Botzmann-Darwin: Redefining Information. Part 2," *Cognitive Semiotics* 2 (2008): 167–194.
6. Ibid.

27. A CONSTRAINT-BASED APPROACH TO EVOLUTIONARY THEORY

1. Stephen Jay Gould, "Sociobiology: The Art of Storytelling," *New Scientist* 80, no. 1129 (1978): 530–533.
2. Terrence Deacon, "A Role for Relaxed Selection in the Evolution of the Language Capacity," *Proceedings of the National Academy of Sciences* 107 (2010): 9000–9006.
3. Terrence Deacon, "Relaxed Selection and the Role of Epigenesis in the Evolution of Language," in *Oxford Handbook of Development Behavioral Neuroscience*, ed. Mark S. Blumberg, John H. Freeman, and Scott R. Robinson (New York: Oxford University Press, 2009), 730–752.
4. Julie Hui and Terrence Deacon, "The Evolution of Altruism via Social Addiction," *Proceedings of the British Academy* 158 (2009): 181–203.
5. Ibid.

28. IMPLICATIONS FOR THE FREE WILL DEBATE?

1. Benjamin Libet, "The Experimental Evidence for Subjective Referral of a Sensory Experience Backwards in Time: Reply to P. S. Churchland," *Philosophy of Science* 48 (1981): 182–197.
2. James W. Haag, Terrence Deacon, and Jay Ogilvy, "The Emergence of Self," in *In Search of Self: Interdisciplinary Perspectives on Personhood*, ed. J. Wentzel van Huyssteen and Erik P. Wiebe, 319–337 (Grand Rapids: Eerdmans, 2011); and James Haig, *Emergent Freedom: Naturalizing Free Will* (Göttingen: Vandenhoeck und Ruprecht, 2008).
3. Ilya Prigogine, *The End of Certainty* (New York: Free Press, 1997).
4. Ibid., 2.
5. Ibid., 5.
6. Ernest Nagel, *The Structure of Science* (Cambridge: Hackett, 1979), conclusion, 312.
7. Douglas Hofstadter, *I Am a Strange Loop* (New York: Basic Books, 2007).
8. Sam Harris, *Free Will Revisited: A Conversation with Daniel Dennett.* Waking Up Podcast by Sam Harris, July 3, 2016.

29. MAKING SCIENCE SAFE FOR VALUE

1. Ursula Goodenough and Terrence Deacon, "Emergence, Ethics, and Religious Naturalism," in *The Oxford Handbook of Religion and Science*, ed. Philip Clayton and Zachary Simpson (Oxford: Oxford University Press, 2006); and Ursula Goodenough and Terrence Deacon, "From Biology to Consciousness to Morality," *Zygon* 38 (2003): 801–819.
2. Stephen Jay Gould, "Nonoverlapping Magisteria," *Natural History* 106 (March 1997): 16–22, www.stephenjaygould.org/library/gould_noma.html.
3. Ernest Becker, *The Denial of Death* (New York: Free Press, 1997).
4. Terrence Deacon, *Incomplete Nature: How Mind Emerged from Matter* (New York: Norton, 2012), 544.
5. George Marshall, *Don't Even Think About It: Why Our Brains Are Wired to Ignore Climate Change* (New York: Bloomsbury USA, 2015).

INDEX